W0036765

A Practical Guide to Compressor Technology

A Practical Guide to Compressor Technology

Editor

Vilas Kulkarni

scitus
academics

A Practical Guide to Compressor Technology

Edited by **Vilas Kulkarni**

Printed in 2017

ISBN: 978-1-68117-325-2

Library of Congress Control Number: 2015939238

© 2016 by
SCITUS Academics LLC,
616, Corporate Way, Suite 2, 4766,
Valley Cottage, NY 10989

www.scitusacademics.com

This book contains information obtained from highly regarded resources. Copyright for individual articles remains with the authors as indicated. All chapters are distributed under the terms of the Creative Commons Attribution License, which permits unrestricted use, distribution, and reproduction in any medium, provided the original author and source are credited.

Notice

Reasonable efforts have been made to publish reliable data and views articulated in the chapters are those of the individual contributors, and not necessarily those of the editors or publishers. Editors or publishers are not responsible for the accuracy of the information in the published chapters or consequences of their use. The publisher believes no responsibility for any damage or grievance to the persons or property arising out of the use of any materials, instructions, methods or thoughts in the book. The editors and the publisher have attempted to trace the copyright holders of all material reproduced in this publication and apologize to copyright holders if permission has not been obtained. If any copyright holder has not been acknowledged, please write to us so we may rectify.

Contents

Preface

Compressors are a vital link in the conversion of raw materials into refined products. Compressors also handle economical use and transformation of energy from one form into another. They are used for the extraction of metals and minerals in mining operations, for the conservation of energy in natural gas reinjection plants, for secondary recovery processes in oil fields, for the utilization of new energy sources such as shale oil and tar sands, for furnishing utility or reaction air, for oxygen and reaction gases in almost any process, for process chemical and petrochemical plants, and for the separation and liquefaction of gases in air separation plants and in LPG and LNG plants. And, as the reader will undoubtedly know, this listing does not even begin to describe the literally hundreds of services that use modern compression equipment.

Editor

Chapter 1

Gas Compressor Station Economic Optimization

Rainer Kurz[1], Matt Lubomirsky[1], and Klaus Brun[2]

[1]Solar Turbines Incorporated, 9330 Skypark Ct., San Diego, CA 92123, USA

[2]Southwest Research Institute, 6220 Culebra Road, San Antonio, TX 78238, USA

ABSTRACT

When considering gas compressor stations for pipeline projects, the economic success of the entire operation depends to a significant extent on the operation of the compressors involved. In this paper, the basic factors contributing to the economics are outlined, with particular emphasis on the interaction between the pipeline and the compressor station. Typical scenarios are described, highlighting the fact that pipeline operation has to take into account variations in load.

INTRODUCTION

The economic success of a gas compression operation depends to a significant extent on the operation of the compressors involved.

Important criteria include first cost, operating cost (especially fuel cost), life cycle cost, and emissions. Decisions about the layout of compressor stations (Figure 1) such as the number of units, standby requirements, type of driver, and type of compressors have an impact on cost, fuel consumption, operational flexibility, emissions, as well as availability of the station.

Figure 1: Typical compressor station with 3 gas turbine driven centrifugal compressors.

CAPITAL COST: FIRST COST AND INSTALLATION COST

Capital cost for a project consists of first cost and installation cost. First cost includes not only the cost for the driver and compressor, and their skid or foundation, but also the necessary systems that are required for operating them, including filters, coolers, instruments, and valves, and, if reciprocating compressors are used, pulsation bottles. Capital spares, operational spares, and start-up and commissioning spares also have to be considered.

Although not intuitively obvious, this is also the area that is affected by driver derates due to site ambient temperature and site elevation: the power demand of the compressor has to be met at site conditions, not at ISO or NEMA conditions.

Installation cost includes all labor and equipment cost to install the equipment on site. It is determined by component weights, as well as the amount of operations necessary to bring the shipped components to working condition.

MAINTENANCE COST

Maintenance cost includes the parts and labor to keep the equipment running at or above a certain power level. This includes routine maintenance (like change of lube oil and spark plugs in gas engines) and overhauls. Maintenance events can be schedule or condition based. A cost related to maintenance effort is the cost due to the unavailability of the equipment (see below). Maintenance affects availability in two ways. Many, but not all, maintenance events require the shutdown of the equipment, thus reducing its availability. Scheduled maintenance has usually less of an effect than unscheduled events. For example, a scheduled overhaul of a gas turbine may keep the equipment out of operation for only a few days if an engine exchange program is available.

On the other hand, insufficient or improper maintenance negatively affects the availability due to more rapid performance degradation and higher chance of unplanned shutdowns.

EFFICIENCY, OPERATING RANGE, AND FUEL COST

The performance parameters of the compressor and its driver that are important for the economic evaluation are efficiency and operating range. Efficiency ultimately means the cost of fuel consumed to bring a certain amount of gas from a suction pressure to a discharge pressure. In technical terms, this would be a package with a high thermal efficiency (or low heat rate) for the driver and a high isentropic efficiency of the compressor, including all parasitic losses (such as devices to dampen pulsations, but also pressure drop due to filtration requirements) combined with low mechanical losses. This factor determines the fuel cost of the unit, operating at given operating conditions.

Operating range describes the range of possible operating conditions in terms of flow and head at an acceptable efficiency, within the power capability of the driver. Of particular importance are the means of controlling the compressor (e.g., speed control for centrifugal machines, or cylinder deactivation, clearance control, and others for a reciprocating machine) and the relationship between head and flow of the system the compressor feeds into. Operating range often determines the capability to take advantage of opportunities to sell more gas. It should be noted that there is no real low flow limit for stations, due to the conceptual capability for station recycle or to shutting down units. Unit shutdown, in turn, has to be considered with regards to starting reliability of the units in question, as well as the impact on maintenance. In this study, operating range and the upside potential are not specifically considered, mostly because no data regarding frequency and value of these upside potentials was available. Upside potential can be realized, if the equipment capability can be used to produce more gas, thus taking advantage of additional market opportunities.

The cost of the fuel gas is not automatically the same as the market price of the transported gas. It depends, among other things, on how fuel usage and transport tariff are related. The cost of fuel is also impacted by whether the operator owns the gas in the pipeline (which makes fuel cost an internal operating cost) or the operator ships someone else's gas. In some instances, the fuel cost might be considered virtually nil. Thus, the ratio between fuel cost and maintenance cost can vary widely. In most installations, however, the fuel cost may account for more than two-thirds of the annual operating cost.

The value of operational flexibility is somewhat hard to quantify, but many pipeline systems operate at conditions that were not foreseen during the project stage. Operational flexibility will result in lower fuel cost under fluctuating operating conditions and in added revenue if it allows shipping more gas than originally envisioned.

EMISSIONS

Any natural-gas-powered combustion engine will produce a number of undesirable combustion products. NOx is the result of the reaction between Nitrogen (in the fuel or combustion air) and oxygen and requires high local temperatures to form. Lean premix gas turbines and lean premix engines reduce the NOx production. Catalytic exhaust gas treatments, such as SCR's, can remove a big portion of NOx in the exhaust gas, but also add ammonia to the exhaust gas stream.

Products of incomplete combustion include VOC's, CO, methane, and formaldehyde. Fuel bound sulfur will form SOx in the combustion process.

The combustion products above are usually regulated. For the economic analysis, the cost of bringing the equipment to meet local or federal limits has to be considered.

CO_2 is the product of burning any type of hydrocarbons. CO_2, and some other gases, such as methane, are considered greenhouse gases. Typically, all greenhouse gases are lumped together into a CO_2 equivalent. In this context, it must be noted that methane is considered about 20 times as potent a greenhouse gas as CO_2. Thus, 1 kg of methane would be considered as about 20 kg CO_2 equivalent. In some countries, CO_2 production is taxed, and there is a possibility that other countries, including the USA, adopt regulations that would give CO_2 avoidance an economic value. In this case, the amount of CO_2 or methane that is released to the atmosphere has to be considered as a cost in the economic evaluation.

It further needs to be considered that the engine exhaust is not the only source of emissions in a compressor station related to the compression equipment. There are also sources of methane leaks in the compression equipment that may have to be considered. In this case, one has to distinguish leakage that is easily captured and can thus be fed to a flare and leakage that cannot be captured easily. Further, it may have to be evaluated how frequently the station has to be blown down. For example, whether the compression equipment can be maintained and started from a pressurized hold determines the amount of unwanted station methane emissions.

Lastly, other consumables may have to be considered. The frequency and cost of lube oil changes, as well as lube oil replacements due to lube oil consumption, generate costs on various levels: first, the replacement cost for lube oil, second, if the lube oil is used in the combustion process, the resulting emissions, and third, if the lube oil enters the pipeline, the cost due to the pipe contamination, including possibly the increased maintenance cost of downstream equipment.

AVAILABILITY

Availability is the ratio between the hours per year when the equipment is supposed to operate and actually can operate and the hours per year where the equipment is supposed to operate. Availability therefore takes into account the entire equipment down time, both due to planned and unplanned maintenance events. In other words, if the operator needs the equipment for 8760 hours per year, and the equipment requires 3 shutdowns, lasting 2 days each for scheduled maintenance, and in addition also had to be shut down for 3 days due to a equipment failure, the availability of the equipment would be 97.5%. Other than MTBF, which describes the frequency of unplanned failures, the availability has a direct impact on the capability of an installation to earn money. Besides the type of equipment, the quality of the maintenance program and the measures taken to deal with environmental conditions (air, fuel, etc.) have a significant effect on the availability (and reliability) of the equipment.

The cost associated with availability is the fact that the station may not be able to perform its full duty for certain amount of time, thus not earning money. The loss of income can be due to the reduction of the pipeline throughput, the unavailability of associated products (oil on an oil platform, condensates in a gas plant), or due to penalties assessed because contractual commitments are not met. The value associated with the lost production is not necessarily the market value of the lost production. It may also be the loss of income from transportation tariffs, the cost from penalties for not being able to satisfy delivery contracts, or the cost for lost opportunity (i.e., due to the requirement to keep spare power to compensate for poor availability, instead of being able to use the spare power to increase contractually agreed deliveries).

Station availability can be improved by installing spare units, but this is an additional first cost factor.

COMPRESSOR OPERATION

The relationship between pressures and flows in any given application that employs gas compressors as well as some other factors may influence the arrangement of compressors in a station as well as the type of equipment used. The question about series or parallel arrangements in a station has to be considered both in the light of steady-state aerodynamic performance, as well as regarding transient behavior, spare strategy, and growth capabilities. Not only has the full load behavior of the driver, but also its part-load characteristics influenced areas like fuel consumption and emissions. Since greenhouse gas emissions from CO_2 for a given fuel gas are only influenced by the overall efficiency of the operation, there is a strong tie between efficiency and emissions.

Different concepts such as the number of units installed, both regarding their impact on the individual station and the overall pipeline and the necessity of standby units are discussed. The number of compressors installed in each compressor station of a pipeline system has a significant impact on the availability, fuel consumption, and capacity of the system. Depending on the load profile of the station, the answers may look different for different applications.

The operating point of a compressor is determined by a balance between available driver power, the compressor characteristic, and the system behavior. The compressor characteristic also includes the means of controlling the compressor, such as

- variable speed control,
- adjustable inlet guide vanes,
- suction or discharge throttling,
- recycling.

If variable speed control is available, for example, because the driver is a two-shaft-gas turbine or a variable speed electric motor, this is usually the preferred control method. It is often augmented by the capability to recycle gas. A typical compressor map for a speed-controlled compressor is shown in Figure (Figure 2). It shows the area of possible compressor operating points. The lowest flow possible is determined by the surge line. If the station requires even lower flows, gas has to be recycled. On any point of the map, compressor speed and power consumption are different.

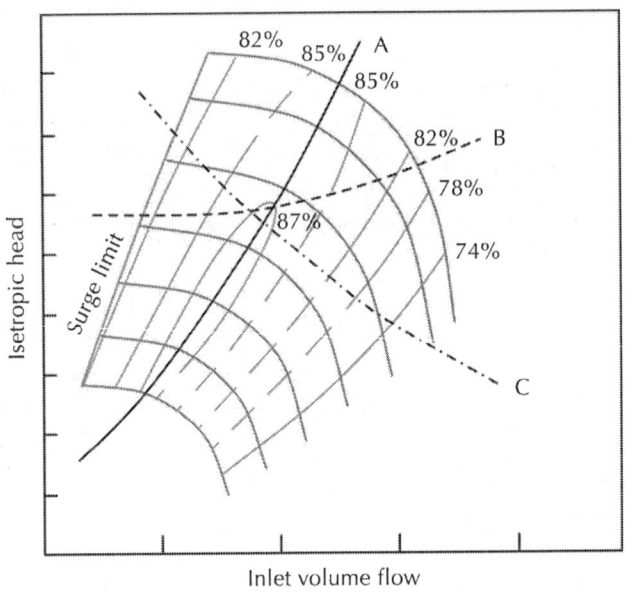

Figure 2: System characteristics and compressor map.

Where on the map the compressor will actually operate is determined by the behavior of the system, that is, the relationship between head (pressure ratio) and flow enforced by the system (Figure 2). Line B depicts a system where suction and discharge pressure are more or less fixed and thus change very little with changes in flow. Examples are refrigeration systems or systems where the suction pressure is fixed by a required separator pressure, and the discharge pressure is fixed by the need to feed that gas into an existing pipeline. Line A shows the typical behavior of a pipeline, where any change in flow will impact the pressure drop due to friction in the pipeline.

Line C is typical for storage applications, where the pressure in the storage cavity increases with the amount of gas stored. If the compressor is operated at maximum power, the initial flow will be high due to the initially low pressure ratio. The more gas is stored in the cavity, the higher its pressure, and thus the required discharge pressure from the compressor becomes. Being power limited, the operating point then moves to a lower flow.

In case of a pipeline, the operating point of the compressor is always determined by the power available from the driver (Figure 3). In the case of a gas turbine driver, the power is controlled by the gas producer speed setting and the pipeline characteristic. We find this point on the compressor map at the intersection between the pipeline characteristic and the available power. Increasing the flow through a pipeline will require more power and more compressor head.

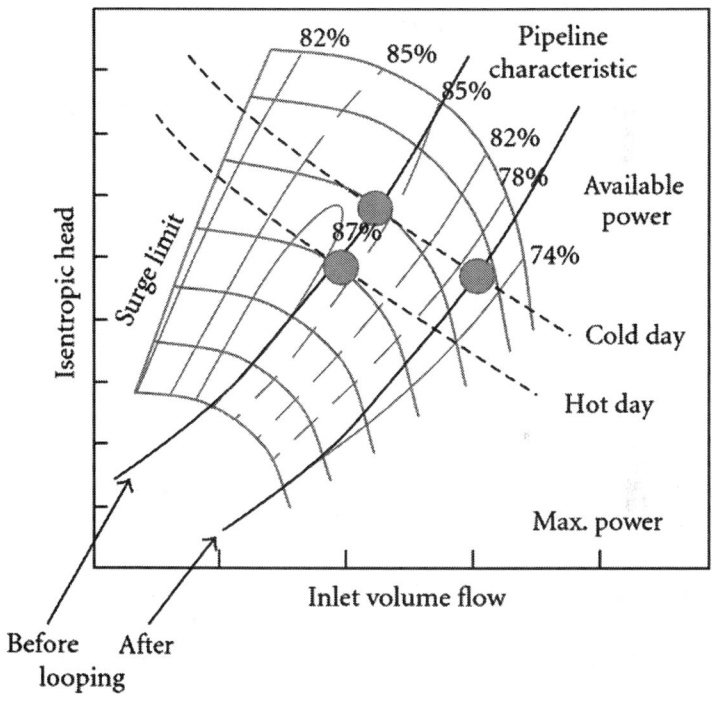

Figure 3: Operating point and driver power.

The change of operating conditions over time is due to many reasons, such as drop in field pressure, the pipeline gets looped or has to meet an increase in capacity. Looping the pipeline will change the characteristic of the pipeline to allow more flow at the same head requirement. So, any change in the pipeline operation will impact power requirements, compressor head or pressure ratio, and flow. Operational changes may move the compressor operating point over time into regimes with lower efficiency. Fortunately, centrifugal compressors driven by gas turbines are very flexible to automatically adapt to these changes to a degree.

As indicated, variable speed drivers allow for efficient operation over a large range of conditions. If a constant speed driver has to be used, the adjustment of the compressor to the required operation usually has to rely on the use of recycling and suction or discharge throttling. Either method, while effective, is not very efficient: throttling requires the compressor to produce more head than required by the process, thus consuming more power. On a compressor speed line, throttling will move the compressor to a point at lower flow than the design point. Recycling forces the compressor to flow more gas than required by the process and thus also consumes more power. It thus allows the compressor to operate at a lower head than the design head. In general, an electric motor for a constant speed drive has to be sized for a larger power than a motor for a variable speed drive under otherwise the same conditions. Furthermore, starting of constant speed motors requires to oversize the electric utilities, especially if a start of a pressurized compressor is desired.

OPERATIONAL FLEXIBILITY AND STANDBY REQUIREMENTS

Operational flexibility under a larger number of different operating scenarios has to be studied. Demand varies on an hourly, daily, monthly, or seasonal basis. Also, the available gas turbine power depends on the prevalent ambient conditions (Lotton and Lubomirsky [1]). Similarly, transient studies on pipeline systems (Santos [2]) can reveal the often large range of operating conditions that needs to be covered by a compressor station and thorough analysis can often reveal which type of concept yields lower fuel consumption. Lastly, scenarios that arise from failures of one or more systems have to be considered (Ohanian and Kurz [3]). In any application, operating limits due to speed limits are usually undesirable, because they mean that the available engine power cannot be used. But, because a gas turbine can produce far more power at colder ambient temperatures, designs based on worst case ambient conditions may not be optimal. Optimization considerations can also be found in [4, 5].

The quest for operational flexibility can be satisfied on various levels: the compressor and the driver should have a wide operating range. Using multiple smaller units per station rather than one large unit can be another way. Here, the arrangement in series or in parallel will impact the flexibility. The gas turbines allow for immediate starting capability if the need arises.

In upstream and midstream applications, configurations usually need to cover a large range of operating points. Depending on the application, the operating points can vary on an hourly, daily, weekly, monthly, or seasonal basis. Contributing factors are supply (e.g., depleted fields or new wells) and demand, changes in gas composition or site available engine power. Often, flow demand and head requirements are coupled. This is very obvious for pipeline applications, where the pressure drop between stations (and thus the pressure ratio of each of the stations) is directly related to the flow (Figure 4).

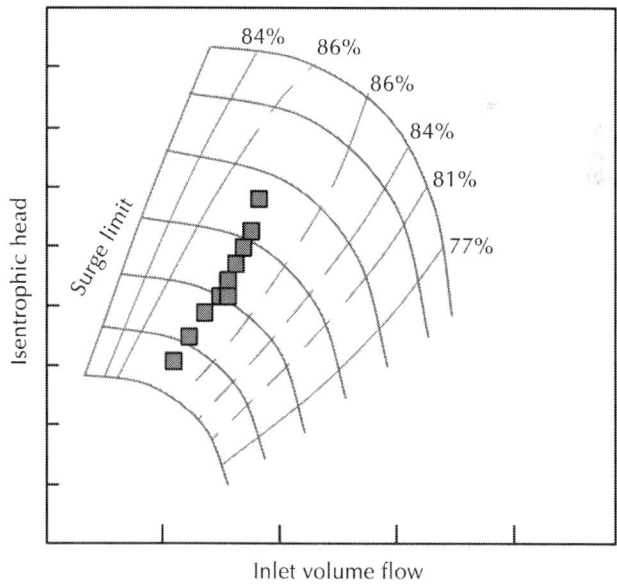

Figure 4: Typical steady state pipeline operating points plotted into a compressor performance map.

In other applications, the compressor operating points are limited by the maximum available engine power. This is, for example, the case for storage and withdrawal operations. Here, the goal is to fill the storage cavity as fast as possible, which means that engine operates at its maximum power. Since the filling of the storage cavity starts at very low pressure differentials, the flow is initially very high. As the cavity pressure and with it the compressor pressure ratio increase, the flow is reduced. For applications like this, compressor arrangements that allow to operate two compressors in parallel during the initial stage, with the capability (either with external or internal valves) to switch to a series operation, are very advantageous. Incidentally, back-to-back machines cannot be used in this case due to their delicate axial thrust balance.

Dynamic studies of pipeline behavior reveal a distinctly different reaction of a pipeline changes in station operating conditions than a steady-state calculation (Figure 5). In steady state (or for slow changes), pipeline hydraulics dictate an increase in station pressure ratio with increased flow, due to the fact that the pipeline pressure losses increase with increased flow through the pipeline (Figure 4). However, if a centrifugal compressor receives more driver power and increases its speed and throughput rapidly, the station pressure ratio will react very slowly to this change. This is due to fact that initially the additional flow has to pack the pipeline (with its considerable volume) until changes in pressure become apparent. Thus, the dynamic change in operating conditions would lead (in the limit case of a very fast change in compressor speed) to a change in flow without a change in head (Figure 5).

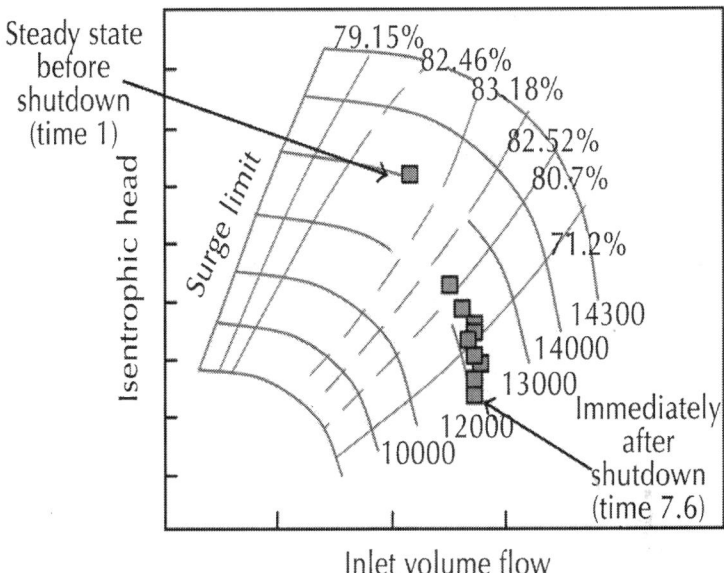

Figure 5: Typical operating points if transient conditions are considered, in this case due to the shutdown of one unit in a two-unit station.

Because the failure or unavailability of compression units can cause significant loss in revenue, the installation of standby units must be considered. These standby units can be arranged such that each compression station has one standby unit, that only some stations have a standby unit, or that the standby function is covered by oversizing the drivers for all stations. (Oversizing naturally creates an efficiency disadvantage during normal operation, when the units would operate in part load.) It must be noted that the failure of a compression unit does not mean that the entire pipeline ceases to operate, but rather that the flow capacity of the pipeline is reduced. Since pipelines have a significant inherent storage capability ("line pack"), a failure of one or more units does not have an immediate impact on the total throughput. Additionally, planned shutdowns due to maintenance can be planned during times when lower capacities are required.

Standby units are not always mandatory because modern gas-turbine-driven compressor sets can achieve an availability of 97% and higher. A station with two operating units and one standby unit thus has a station availability of $100 (1-0.03^2)=99.91\%$(because two units have to fail at the same time in order to reduce the station throughput to 50%). A station with one standby unit and one operating unit also yields a 99.91% station availability. However, while failure of two units in the first case still leaves the station with 50% capacity, the entire station is lost if both units fail in the second case. Arguably, installing two smaller 50% units rather than one larger 100% unit could avoid the need for installing a standby unit.

It has often been assumed that for two-unit stations without a standby unit, a parallel installation of the two units would yield the best behavior if one unit fails. However, Ohanian and Kurz [3] have shown that a series arrangement of identical compressor sets can yield a lower deficiency in flow than a parallel installation. This is due to the fact that pipeline hydraulics dictate a relationship between the flow through the pipeline and the necessary pressure ratio at the compressor station. For parallel units, the failure of one unit forces the remaining unit to operate at or near choke, with a very low efficiency. Identical units in series, upon the failure of one unit, would initially require the surge valve to open, but the remaining unit would soon be able to operate at a good efficiency, thus maintaining a higher flow than in the parallel scenario. Given the fact that the gas stored in the pipeline will help to maintain the flow to the users, a series installation would often allow for sufficient time to resolve the problem.

Operating multiple units (either in parallel or in series) can be optimized on the station or unit level by load sharing controls. If the units are fairly similar in efficiency and size, control schemes that share load such that both units operate at the same surge margin of the compressors can be advantageous and will usually result in a good overall efficiency. Similar (or identical) units in general achieve the lowest overall fuel consumption if both are about evenly loaded. The fuel savings from running one unit at high load (and thus higher gas turbine efficiency) is more than compensated by running the other unit at low load and lower efficiency. In other words, running two units at 75% load results in lower overall fuel consumption than running one unit at 100% and one unit at 50% load.

On the other hand, if the units are dissimilar in size, or of very different efficiency, it may be best if the larger, or the more efficient, unit carries the base load, while the smaller, or the less efficient, unit is responsible to provide power for load swings.

Pipelines with load swings (Figure 6) can often benefit from using multiple smaller units as opposed to single big units.

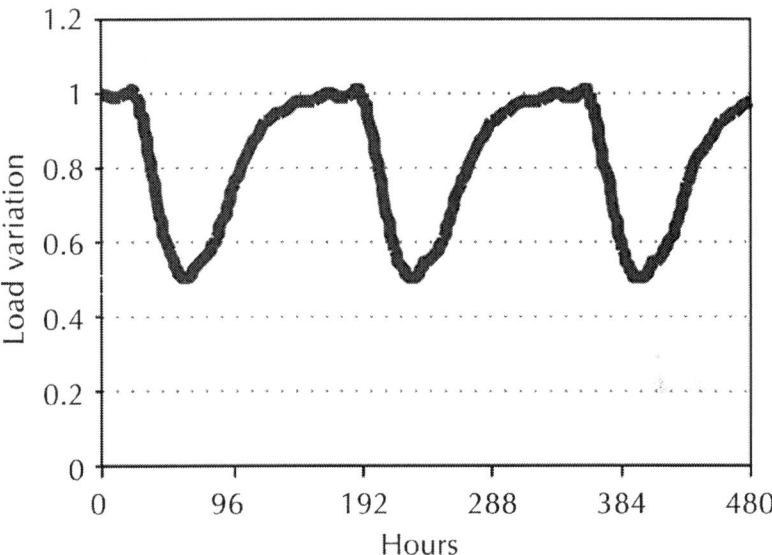

Figure 6: Averaged load variation for four stations of an interstate pipeline in South America during summer and winter scenarios.

While analyzing the performance of the entire package it is important to understand how to distinguish the unit with the best overall efficiency. The turbo compressor performance depends on efficiency of two main components—gas turbine and the centrifugal compressor. The best efficiency compressor does not always provide the overall lowest unit fuel consumptions as the very important piece of the equation is the turbine efficiency. Also, the relationship between the compressor running speed and the power turbine optimum speed at the required given compressor load must be considered. The main goal during the compressor selections process is to find the stage selections that not only provides the highest compressor efficiency but also would yield the highest overall package efficiency, which is achieved, when the fuel consumption for the duty is minimized. As we know from diagram in Figure 7 the lower the turbine part load is-the lower the turbine efficiency will be. As such, the higher loaded turbine should generally provide the better turbine and overall package efficiency.

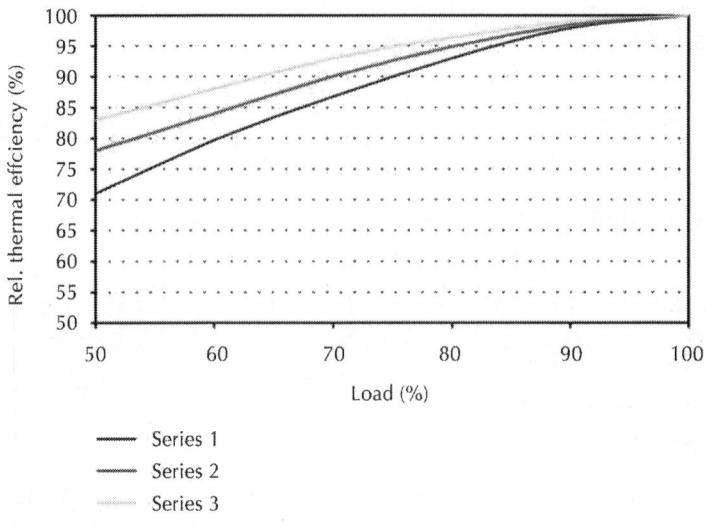

Figure 7: Typical Change of Efficiency with part load for 3 different industrial gas turbines.

The relationship between compressor efficiency and fuel consumption can mathematically be derived as follows.

For a given operation requirement, defined by standard flow, suction pressure, suction temperature, gas composition, and discharge temperature, this operation requirement precisely defines the isentropic head H_s and the mass flow W.

The power consumption of the compressor P_{compr} then only depends on its isentropic efficiency η_s and the mechanical losses P_{mech}, as follows:

$$P_{compr} = W \cdot \frac{H_s}{\eta_s} + P_{mech} \cdot \qquad (1)$$

In other words, for a given duty, the compressor with the better efficiency will always yield lower power consumption, since the mechanical losses tend to be very similar for different compressors.

P_{compr} is the power that must be produced by the gas turbine. In turn, the gas turbine efficiency is a function of the relative load, that is, what percentage of gas turbine full power $P_{GT,max}$ available at the given ambient conditions and the required compressor speed is required by the centrifugal compressor (Kurz and Ohanian [6]):

$$\text{Load}(\%) = \frac{P_{compr}}{P_{GT,max}} \cdot 100.$$

$$(2)$$

The relationship between gas turbine power, fuel flow FF, heat rate HR, and gas turbine efficiency η_{GT} is as follows:

$$FF = P_{GT} \cdot HR = \frac{P_{GT}}{\eta_{GT}}.$$

$$(3)$$

For the compressor applications this means that, for a better compressor, the reduced power consumption indeed causes a small increase in gas turbine heat rate or reduction in gas turbine efficiency. However, the result of lower load and higher heat rate is always lower fuel consumption.

Other important issues must be considered when analyzing overall package efficiency. In many instances it has been requested by the end user to use the gas turbine heat rate as an indication of the overall package performance efficiencies and as the basis for the package guarantees. Higher turbine load will correlate with higher turbine efficiency. This relationship is correct however; the turbo compressor user should recognize some underlying circumstances that can lead to the wrong conclusion.

The peculiar thing is that operating at lower compressor efficiency the centrifugal compressor will require more compression power to do their duty that will ultimately increase the turbine load. If we account that higher part load will lead to the better engine efficiency we discover that compressor with lower efficiency will force turbine operate at better heat rate! This is the fact that is being overlooked in many instances if the comparison between the vendors is done solely based on driver's Heat Rate. The question is how to avoid this trap. The simple and the straight forward answer is that overall unit comparison should be done not on turbine's heat rate, which is simply just a number, but rather on direct value-turbine fuel consumption. In this case the lower compressor efficiency will drive higher turbine power, and, despite the fact that the turbine's heat rate and efficiency will be improving, the actual turbine's fuel consumption will be going up.

In the end, the only measure that is taken into account when calculating package overall operating cost (OPEX) is the fuel consumption whereas the turbine's heat rate or the turbine's simple cycle efficiency remains only numbers on the paper (Figure 8).

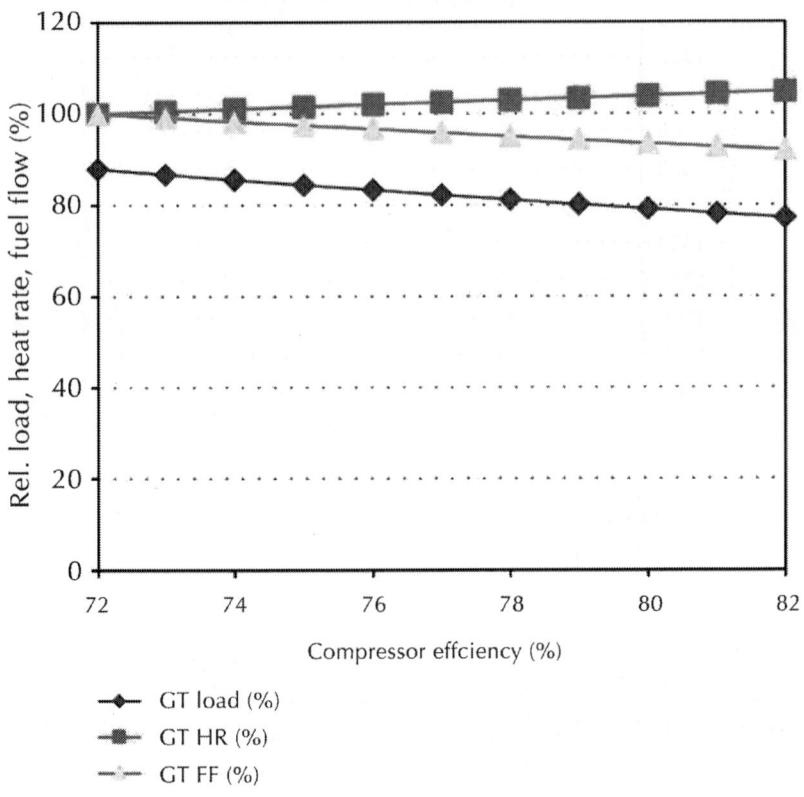

Figure 8: Impact of compressor efficiency on GT Load, heat rate (HR), and fuel consumption (FF).

PIPELINE SIZING CONSIDERATIONS

Kurz et al. [7] evaluated different options for pipe diameters, pressure ratings, and station spacing for a long distance pipeline. A 3220 km (2000 mile) onshore buried gas transmission pipeline for transporting natural gas with a gravity of 0.6 was assumed (Figure 9).

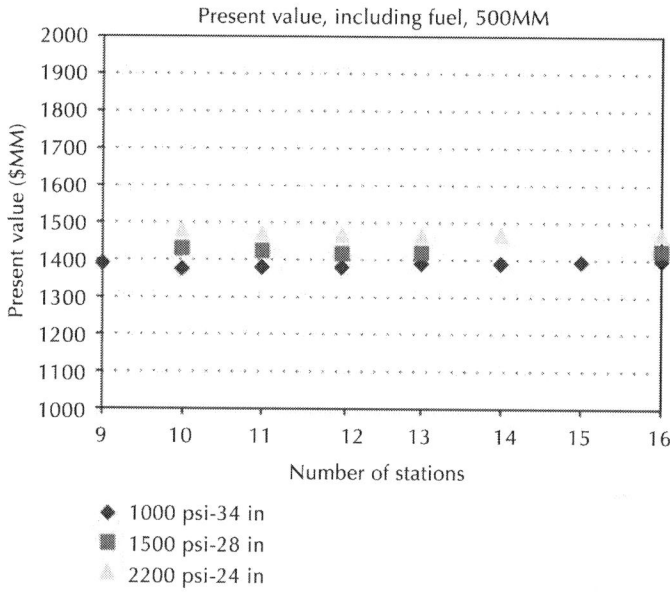

Figure 9: Optimum number of stations and optimum maximum operating pressure (MAOP) for the 3220 km (2000 mile), 560000 Nm³/h sample pipeline. The lowest cost configurations for each MAOP solution are marked (from [7]).

Assuming that pipes will be available in 2'-diameter increments from pipe mills, the nearest even increments of the above-mentioned theoretical diameters were selected (24', 28', and 34' for 152 bar, 103 bar, and 69 bar (2200, 1500, 1000 psia) pressures, resp.) and analyzed by varying the number of stations along the pipeline. The result of this refinement is shown in Figure 9, where present value is plotted against number of stations for each pressure level. The minima for each is shown in the chart with present value total horsepower, and number of stations. In this study, the 69 bar (1000 psia) pipeline has the lowest present value thus would be the most cost-effective solution.

In actual practice, for commonality reasons, identical size units will be installed in the stations. In order to have identical power at each station, the station spacing will be adjusted (dependent on the geography) since the stations at the beginning of the line will consume more power than the stations at the end of the line due to the power required for fuel compression. Identical power at each location also depends on site elevation and design ambient temperature, which would define the site available power from a certain engine.

One of the key findings is that the optimum is relatively flat in all cases. This means in particular that certain considerations may favor larger station spacing, with higher station pressure ratios and higher MAOP in situations where pipelines are routed through largely unpopulated areas.

TYPICAL APPLICATION

For a case study we consider an international long distance pipeline. The total length of the line is about 7000 km. The pipeline consists of two 42' parallel lines which turn into single a 48' line at the crossing of an international border. The pipeline design throughput is 30 billion Nm^3 per year and maximum operating pressure of this pipeline is 9.8 MPa. There are 10 compressor stations planned in one area and over 20 stations in the receiving country. After first gas, it takes 5 years to build up to full capacity.

When we compare operations of the compressor station we need to recognize two main approaches. We can either operate with fewer of larger turbo compressor units (Case A, 2 large units) or with a higher quantity of smaller turbo compressor units (Case B, 4 small units). The following factors need to be considered when selection of either option is decided. In evaluating the system reliability and maximum throughput the impact of unit outages needs to be considered. If we were to consider two large 30 MW units the failure of one of them will result in 50% reduction of power available whereas if we consider 4 smaller 15 MW units the failure of one of them will result in only 25% power reduction. Figure 10 outlines the basic fact that, if the surviving units run at full load to make up as much flow as possible, the operating point for the Case B will be close to the highest efficiency island so the remaining on-line compressors will be working more effectively compared to Case A when the single remaining large unit will be working in the stonewall area. It is obvious that pipeline recovery time will be shorter in Case B.

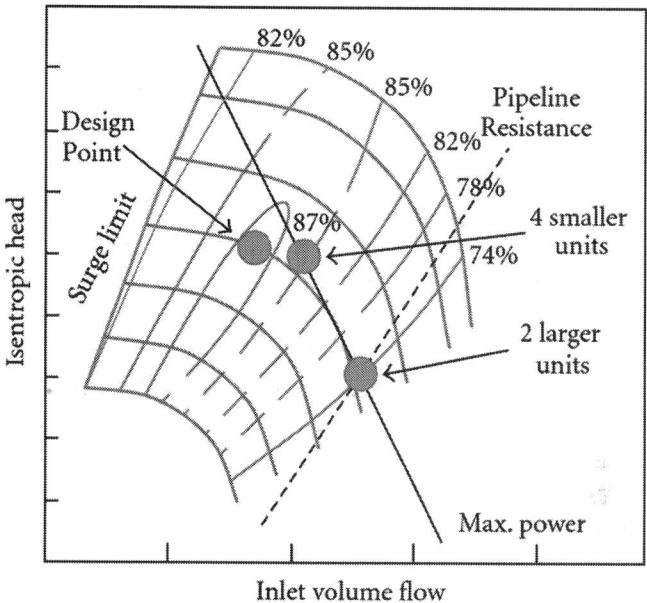

Figure 10: Impact of loss of one unit for the 4 unit and the 2 unit scenarios.

Based on an analysis by Santos et al. [8, 9], Case A can represent even more problems. The amount of gas that the single remaining 30 MW unit will have to process is so big that it will put this remaining unit into choke, and thus for practical purposes out of operation. The amount of fuel that the remaining unit is going to burn will not justify that negligible increase in head that this unit will provide. So, practically, when one larger turbocompressor will be out of operation, the second will have to be shut down and the station will be bypassed. Station configurations with the single oversize driver and either no standby or standby on each second or third station are often advocated. The arguments in favor of this method are very high pipeline availability (99.5%) and high efficiency (40–42%) of the larger 30 MW turbocompressor units. In fact, designing for a turbine oversized by 15% will lead to normal operations at part load conditions almost all the time (99.5%) where there will be negative impact on turbine efficiency and, as a result of it, increased fuel consumption Another negative impact of this approach is that normally this pipeline would operate at lower than MAOP pressure, whereas the highest operational pipeline pressure produces less pressure losses and, therefore, lower requirements for the recompression power. The reason for that is the maintenance schedule for the turbocompressors on the stations with the single units without standby. In order to perform maintenance on these units the pipeline, linepack will have to be maximized up to MAOP, so that unit can be taken off line and the pipeline throughput will not be impacted. Therefore, the normal pipeline operations have to be based on a lower MAOP. Also worth mentioning is the pipeline capacity when considering the single turbocompressor approach. Many pipelines transport gas owned and produced by different commercial entities. As such the gas fields development time and gas availability depend on many technological and, lately, political factors that may potentially have negative impact on pipeline predicted capacity growth. In these conditions the single oversized turbocompressor will either be working into deep recycling mode until the expected amount of gas will become available or start operation with smaller capacity compressor stages which will subsequently require a costly change of the internal bundle.

FUEL COMPARISON

It is increasingly important to evaluate all seasons conditions when making a comparison between two different station layout cases. For the subject pipeline different design organizations were involved in the pipeline feasibility study. One of them has used summer conditions only and came to the conclusion that larger turbines are preferred option. Another source used annual average conditions and came to the opposite conclusion. The reason for that was the fact that during winter, fall, and spring months, which cover total of 9 out of 12 months of operation, one of the smaller turbo compressors was put in the standby mode. Due to lower ambient temperature the amount of power available from the remaining 3 units was enough to cover the 100% duties due to high compressor efficiency. This was not true for Case 1 (based on same explanation above) and both 30 MW units had to work in the deep part load with unsatisfactory turbine efficiency. The fact that operational mode became 3 + 2 for Case B gave additional benefits worth mentioning. Since two turbocompressors were in standby mode there was an opportunity to do all maintenance work during this time of the year. It means that availability of this system becomes superior compared to Case A, especially if we were to consider summer months of operations.

MAINTENANCE AND OVERHAULS

Another advantage of operating only 3 out of 5 units for a significant part of the year (i.e., 9 out of 12 months) is the extended time between overhauls. Based on the calculations below, the total number of hours for each turbocompressor unit per year was reduced from 7008 to 5694, and, therefore, the time between overhauls could be extended. Based on 4+1 units operating during 3 summer months and 3 + 2 units operating during the rest of the year (9 months), if the units were used so that they all ran exactly the same number of hours each year, each unit would run for 5,694 hours every year. Whereas if we account for 4 working units with one standby throughout the year the number of working hours will be as follows: 8,760 × 4 units running = 35,040/5 units available = 7,008 total hours per unit/year.

Note that all units run for an equal number of hours to make the calculation simple. However, the customer could push lead machines to reach the agreed time between major inspections (TBI) first, so that all engines do not come up for overhaul at the same time, and this would help with the overhaul cost, helping to distribute the overhaul cost over the 30 year cycle. We can even make step further and will see additional benefits of this approach. Accounting for the normal year around operation with 4 units online, each turbocompressor will get 7,008 × 30 year = 210,240 required hours of operations, whereas considering 3 + 2 setup for 9 months the total number of the required hours of operations reduced down to 170,820 hours. With modern turbines technology it is not uncommon to see that lifetime operation reaches 150–180,000 hours. It means that for the lifetime of this project (30 years) there will be no need to buy new set of equipment. This alone makes huge favorable impact on projects economics.

STATION VERSUS SYSTEM AVAILABILITY

It is important to recognize the difference between station and pipeline availability. For economic assessments, misunderstanding this issue can lead to the wrong conclusion. Station availability calculations are easy, straight forward and based on simple statistical equations. It is easy to see that fewer units on a compressor station will yield higher availability, assuming the threshold for availability is 100% of the flow. But is this true for the entire pipeline system? The answer is not easy and requires additional investigation including extensive hydrodynamic analysis using of the statistical methodology. The Monte Carlo method [9] has proved to be the good methodology to determine the pipeline system availability. The statistical portion consists of generating multiple random cases of equipment failure on single or two consecutive compressor stations. The hydrodynamic portion will calculate the maximum throughput that pipeline is available to carry when these failures occur. Based on this extensive and in-depth analysis it can be shown that availability of a pipeline, configured with smaller multiple units, delivers better overall results. The main reason for that outcome is the fact that shutdown of the smaller unit makes lesser impact on the behavior of the entire pipeline. Of course, to have fair results, the availability of the single turbocompressor unit, either smaller 15 MW or larger size 30 MW, was identical. It is easy to understand that in our particular case the availability of the station setup with smaller units (Case B) was greatly enhanced because of the presence of extra standby unit during winter and fall/spring months when stations setup has 3 + 2 configuration.

EFFECT OF LARGE UNIT SHUTDOWN

Examples of the vulnerability are demonstrated based on a typical pipeline scenario with 4 stations. Each station has 2 compressor trains without spares. If one unit in station 2 is lost, the pipeline flow is reduced by 12%. However, the same 12% flow reduction can be maintained by also shutting down the surviving unit in station 2. This is due to the necessarily inefficient operation of the surviving unit in station 2, which is forced to operate in choke. If both units are shutdown, stations 3 and 4 will be able to recover the flow, but at a much higher overall efficiency. Thus, shutting both units down reduces the pipeline fuel consumption compared to the scenario with only one unit shut down in station 2. The point of this example is, the failure of one of two large units in a compressor station has more significant consequences than the failure of a smaller unit in a station with three or more operating units. Or, in other words, scenarios with 3 or more units per station without spare units tend to have a higher flow if one of the units fails or has to be shut down for maintenance, than scenarios with 2 units per station without spare units.

CONCLUSIONS

The paper has illustrated the different influence factors for the economic success of a gas compression operation. Important criteria include first cost, operating cost (especially fuel cost), capacity, availability, life cycle cost, and emissions. Decisions about the layout of compressor stations such as the number of units, standby requirements, type of driver and type of compressors have an impact on cost, fuel consumption, operational flexibility, emissions, as well as availability of the station.

REFERENCES

1.   J. H. Lotton and M. Lubomirsky, Gas Turbine Driver Sizing and Related Considerations, Bratislava Publishers, Slovakia, 2004.

2.   S. P. Santos, "Transient analysis- a must in gas pipeline design," in Proceedings of the Pipeline Simulation Interest Group Conference, Tucson, Ariz, USA, 1997.

3.   S. Ohanian and R. Kurz, "Series of parallel arrangement in a two-unit compressor station," Journal of Engineering for Gas Turbines and Power, vol. 124, no. 4, pp. 936–941, 2002. · ·

4.   W. Wright, M. Somani, and C. Ditzel, "Compressor station optimization," in Proceedings of the 30th Annual Meeting, Pipeline Simulation Interest Group, Denver, Colo, USA, 1998.

5.   M. Pinelli, A. Mazzi, and G. Russo, "Arrangement and optimization or turbocompressors in off-shore natural gas extractions station," Tech. Rep. GT20005-68031, ASME, 2005.

6.   R. Kurz and S. Ohanian, "Modelling turbomachinery in pipeline simulations," in Proceedings of the 35th Annual Meeting, Pipeline Simulation Interest Group, Berne, Switzerland, 2003.

7.   R. Kurz, S. Ohanian, and K. Brun, "Compressors in high pressure pipeline applications," Tech. Rep. GT2010-22018, ASME, 2010.

8.   S. P. Dos Santos, M. A. S. Bittencourt, and L. D. Vasconcellos, "Compressor station availability—managing its effects on gas pipeline operation," in Proceedings of the Biennial International Pipeline Conference (IPC ‹07), vol. 1, pp. 855–863, 2007.

9.   S. P. Santos, "Monte Carlo simulation—a key for a feasible pipeline design," in Proceedings of the Pipeline Simulation Interest Group Conference, Galveston, Tex, USA, 2009.

DCMS: A Data Analytics and Management System for Molecular Simulation

Anand Kumar[1], Vladimir Grupcev[1], Meryem
Berrada[1], Joseph C Fogarty[2], Yi-Cheng Tu[1],
Xingquan Zhu[3], Sagar A Pandit[2], and Yuni Xia[4]

[1]Department of Computer Science and Engineering, University of South Florida, 4202 E. Fowler Ave., ENB118, Tampa 33620, Florida, USA

[2]Department of Physics, University of South Florida, 4202 E. Fowler Ave., PHY114, Tampa 33620, Florida, USA

[3]Department of Electrical Engineering and Computer Science, Florida Atlantic University, 777 Glades Road, EE308, Boca Raton 33431, Florida, USA

[4]Department of Computer Science, Indiana University - Purdue University Indianapolis, 723 W. Michigan St, SL280E, Indianapolis 46202, Indiana, USA

ABSTRACT

Molecular Simulation (MS) is a powerful tool for studying physical/ chemical features of large systems and has seen applications in many scientific and engineering domains. During the simulation process, the experiments generate a very large number of atoms and intend to observe their spatial and temporal relationships for scientific analysis. The sheer data volumes and their intensive interactions impose significant challenges for data accessing, managing, and analysis. To date, existing MS software systems fall short on storage and handling of MS data, mainly because of the missing of a platform to support applications that involve intensive data access and analytical process. In this paper, we present the database-centric molecular simulation (DCMS) system our team developed in the past few years. The main idea behind DCMS is to store MS data in a relational database management system (DBMS) to take advantage of the declarative query interface (i.e., SQL), data access methods, query processing, and optimization mechanisms of modern DBMSs. A unique challenge is to handle the analytical queries that are often compute-intensive. For that, we developed novel indexing and query processing strategies (including algorithms running on modern co-processors) as integrated components of the DBMS. As a result, researchers can upload and analyze their data using efficient functions implemented inside the DBMS. Index structures are generated to store analysis results that may be interesting to other users, so that the results are readily available without duplicating the analysis. We have developed a prototype of DCMS based on the Postgre SQL system and experiments using real MS data and workload show that DCMS significantly outperforms existing MS software systems. We also used it as a platform to test other data management issues such as security and compression.

BACKGROUND

Recent advancement in computing and networking technologies has witnessed the rising and flourishing of data-intensive applications that severely challenge the existing data management and computing systems. In a narrow sense, data-intensive applications commonly require significant storage space and intensive computing power. The demand of such resources alone, however, is not the only fundamental challenge of dealing with big data [1]-[3]. Instead, the complications of big data are mainly driven by the complexity and the variety of the data generated from different domains [4]. For example, online social media has now been popularly used to collect real-time public feedback related to specific topics or products [5]. Data storage and management systems should support high throughput data access with millions of tweets generated each second [4]. Meanwhile, the tweets may be generated from different geographic regions, using different languages, and many of them may contain spam messages, typos, and malicious links etc. In addition to the low level data cleansing, access, and management issues, user privacy and public policies should also be considered (and integrated) in the analytical process for meaningful outcomes.

For many other application domains, such as scientific data analysis, the above big data complications also commonly exist. For example, particle simulation is a major computational method in many scientific and engineering fields for studying physical/chemical features of natural systems. In such simulations, a system of interest is treated as a collection of potentially large number of particles (e.g., atoms, stars) that interact under classical physics rules. In the molecular and structural biology world, such simulations are generally called Molecular Simulations (MS). By providing a model description for biochemical and biophysical processes at a nanoscopic scale, MS is a powerful tool towards fundamental understanding of biological systems.

At present time, the field of MS has a handful of software systems employing their proprietary or open formats for data storage [6]-[8]. Although many of them are carefully designed to achieve maximum computational performance in simulation, they significantly fall short on storage and handling of the large scale data output. The MS by their nature generate a large amount of data in a streaming fashion - a system could consist of millions of atoms and one single simulation can easily run for tens of thousands of time steps. Figure 1 shows two (small) examples of such simulations. One salient problem of existing systems is the lack of efficient data retrieval and analytical query processing mechanisms.

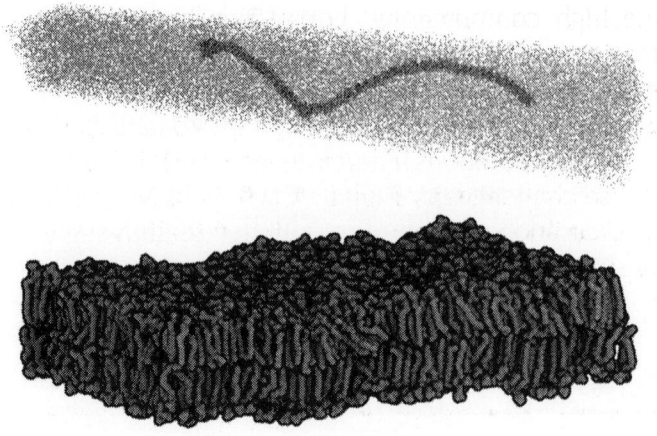

Figure 1: Snapshots of two MS systems: a collagen fiber structure with 890,000 atoms (top) and a dipalmitoylphosphatidylcholine (DPPC) bi-layer lipid system with 402,400 atoms (bottom).

In this paper, we present our recent research efforts in advancing big data analytic and management systems for scientific simulation domains, which usually generate large datasets with temporal and spatial correlations for analysis. Our research mainly emphasizes on the design of the data management system in supporting intensive data access, query processing, and optimization mechanisms for MS data. The main objective of our study is to produce high performance techniques for the MS community to accelerate the discovery process in biological/medical research. In particular, we introduce the design and development of a Database-Centric Molecular Simulation (DCMS) framework that allows scientists to efficiently retrieve, query, analyze, and share MS data.

A unique feature of DCMS is to build the system framework on top of a relational database management system (RDBMS). Such a decision is justified by careful analysis of the data processing requirements of the target application: since MS data is spatiotemporal in nature, existing DBMS provides significant advantages in modeling and application development. Plus, we can leverage the results of decades of research in spatiotemporal databases that are often (at least partially) implemented in RDBMSs. On the other hand, the unique features of MS data analysis/querying workload call for significant improvement and new functionalities in existing RDBMSs. A salient problem we face is the high computational complexity in processing analytical queries that are not seen in typical databases, demonstrating another dimension of difficulty shared by many of today's big data applications. For MS, there are also data compression and data security issues that require innovative solutions. Therefore, our work in DCMS focuses on meeting those challenges by augmenting the DBMS kernel with novel data structures and query processing/optimization algorithms. As a result, our system achieves significant improvement in data processing efficiency as compared to legacy solutions.

Related Work

In current MS software [9]-[11], simulation data is typically stored in data files, which are further organized into various levels of directories. Data access is enabled by encoding the descriptions of the content in files into the names of files and directories, or storing more detailed descriptions about the file content in separate *metadata* files. Under the traditional file-based scheme, data/information sharing among MS community involves shipping the raw data packed in files along with the required format information and analysis tools. Due to the sheer volume of MS data, such sharing is extremely difficult, if possible at all. Two MS data analysis projects, BioSimGrid [12] and SimDB [13], store data and perform analysis at the same computer system and allow users remotely send in queries and get back results. This approach is based on the premises that: (1) analysis of MS data involves projection and/or reduction of data to smaller volume; (2) users need to exchange the reduced representation of data, rather than the whole raw data. In a similar project [14], databases are used to store digital movies generated from visualization of MS datasets.

In BioSimGrid and SimDB, relational databases are used to store and manage the metadata information. However, both systems *store raw MS data as flat files instead of database records*. Thus, the database only helps in locating the files of interest by querying the metadata. Further data retrieval and analysis are performed by opening and scanning the files located. Such an approach suffers from the following drawbacks: (1) *Difficulties in the development and maintenance of application programs*. Specific programs have to be coded for each specific type of queries using a general-purpose programming language such as C. This creates high demand for experienced programmers and thus limits the type of queries the system can support. (2) *Lack of a systematic scheme for efficient data retrieval and analysis.* An operating system views data as continuous bytes and only provides simple data access interfaces such as *seek* (i.e., jumping to a specific position of the file). Without data structures that semantically organize data records, data retrieval is often accomplished by sequentially scanning all relevant files. There is also a lack of efficient algorithms for processing queries that are often analytical in nature - most of existing algorithms are brute-force solutions. (3) *Other issues* such as data security and data compression are not sufficiently addressed.

The MDDB system [8] is close in spirit to DCMS. However, it focuses on data exploration and analysis within the simulation process rather than post-simulation data management. Another project named Dynameomics [15] coincided with the development of DCMS and delivered a database containing data from 11,000 protein simulations. Note that the main objective of the DCMS project is to provide a systematic solution to the problems mentioned above. To that end, most of our work is done within the kernel space of an open-source DBMS. In contrast to that, Dynameomics uses a commercial DBMS in its current form and attempts to optimize data management tasks at the application layer. We believe the DCMS approach has significant advantages in solving the last two issues mentioned above.

CASE DESCRIPTION

Issues

Here we summarize the data management challenges in typical MS applications.

MS Data: A typical simulation outputs multiple *trajectory files* containing a number of snapshots (named *frames*) of the simulated system. Depending on the software and format, such data may be stored in binary form and undergo simple lossless compression. The main part of the data is very similar to those found in spatiotemporal databases. A typical trajectory file has some global data, which is used to identify the simulation, and a set of frames arranged in a sequential manner. Each frame may contain data entries that are independent of the atom index. The main part of trajectory frame is a sequential list of atoms with their positions, velocities, perhaps forces, masses, and types. These entries may contain additional quantities like identifiers to place an atom in particular residue or molecule. In file-based approach, the bond structure of residues is stored separately in *topology* files and the control parameters of a simulation are kept separately in *control* files. Hence, any sharing of data or analysis requires consistent exchange or availability of three types of files. Further complications in data exchange/use is due to different naming and storage convention used by individual researchers.

MS Queries: Unlike traditional DBMSs where data retrieval is the main task, the mainstream queries in DCMS are analytical in nature. In general, an analytical query in MS is a mathematical function that maps the readings of a group of atoms to a scalar, vector, a matrix, or a data cube [13]. For the purpose of studying the statistical feature of the system, popular queries in this category include density, first-order statistics, second-order statistics, and histograms. Conceptually, to process such queries, we first need to retrieve the group of atoms of interest, and then compute the mathematical function. Current MS analysis toolboxes [6], [7], [9], [11] accomplish these steps in an (algorithmically) straightforward way. Some of the analytical queries are computationally expensive. Popular queries can be found in Table 1 and we will elaborate more on those in Section "Analytical queries in DCMS". Many types of analytical queries are unique to the MS field, especially those that require the counting of all *n*-body interactions (thus named *n-body correlation functions*). For example, the spatial distance histogram (SDH) is a 2-body correlation function in which all pairwise distances are to be counted. The query processing engines in traditional DBMSs are designed with only simple aggregate queries in mind. Therefore, one major challenge of this project is to design mechanisms for efficient processing of analytical queries in MS that go far beyond simple aggregates.

Table 1: Popular analytical queries in MS

Query name	Definition/Description
Moment of inertia	$I = \sum\limits_{i-1}^{n} m_i \mathbf{r}_i$
Moment of inertia on z axis	$I_z = \sum\limits_{i-1}^{n} m_i r_{zi}$
Sum of masses	$M = \sum\limits_{i-1}^{n} m_i$
Center of mass	$C = \frac{I}{M}$
Radius of gyration	$RG = \sqrt{\frac{I_z}{M}}$
Dipole moment	$D = \sum\limits_{i-1}^{n} q_i \mathbf{r}_i$
Dipole histogram	$D_z = \sum\limits_{i-1}^{n} \frac{D}{z}$
Electron density	$ED = \dfrac{\sum\limits_{i=1}^{n}(e_i - q_i)}{d_{zz} \ast x \ast y}$
Heat capacity	$\dfrac{3000\sqrt{T} \ast bolt_z}{2 \ast \sqrt{T} - n \ast df \ast VarT}$
Isothermal compressibility	$\dfrac{VarV}{V_{avg} \ast bolt_{zz} \ast T \ast PresFac}$
Mean square displacement	$msd = \langle (r_{t+\Delta_t} - r_t) \rangle$

Diffusion constant	$D_t = \frac{6*msd(t)}{t}$
Velocity autocorrelation	$V_{acor} = \langle\langle (V_{t+\Delta_t} * V_t) \rangle\rangle$
Force autocorrelation	$F_{acor} = \langle\langle (F_{t+\Delta_t} * F_t) \rangle\rangle$
Density function	Histogram of atom counts
Spatial distance histogram (SDH)	Histogram of all atom-to-atom distances
RDF	$rdf(r) = \frac{SDH(r)}{4*\pi*r^2*\sigma_r*\rho}$

Kumar*et al.*

Kumar*et al.* *Journal of Big Data* 2014 **2**:9 doi: 10.1186/s40537-014-0009-5

While analytical queries are the workhorse tools for scientific discovery, many require retrieval of data as the first step. Furthermore, visualization tools also interact with the database retrieving subsets of the data points. By studying the data access patterns of the analytical queries and such tools, we identify the following data access queries that are relevant in DCMS:

Point queries are equivalent to accessing a single point at the 3D space, e.g., find the location and/or other physical measurements of an atom at a specific time frame. Such queries are extremely useful for many visualization tools. A typical scenario is: the visualization tool asks for a sample of n data points within a specific region that reflects the underlying distribution. This is done by issuing queries with randomly generated atom IDs.

Trajectory queries retrieve all data points by fixing the value in one dimension. Two queries in this category are very popular in MS analysis: (1) single-atom trajectory (TRJ) query that retrieves the readings of a specific atom along time, and (2) frame (FRM) query that asks for the readings of all atoms in a specific time frame. These two queries, especially the second, are often issued to retrieve data points for various analytical queries such as the diffusion coefficient, in which we compute the root mean square displacement of all atoms in a frame.

Range (RNG) queries are generalized trajectory queries with range predicates on one or more dimensions. For example, find all atoms in a specific region of the simulated space, or, find all atoms with velocity greater than 50 and smaller than 75. Range queries are the main building blocks of many analytical queries and visualization tasks.

Nearest neighbor (NN) queries ask for the point(s) in a multidimensional space that are closest to a given point. For example, *retrieve the 20 closest atoms to a given iron atom.* This may help us locate unique structural features, e.g., certain part of the protein where a metal ion is bound to.

DCMS Architecture

The architecture of the DCMS framework is illustrated in Figure 2, where the solid lines represent command flow and dotted lines represent data flow. At the core of DCMS is an integrated database system, including simulation parameters/states, simulation output data, and metadata for efficient data retrieval. An important design goal of DCMS is to allow scientists to easily transfer, interrogate, visualize, and hypothesize from integrated information obtained from a user-friendly interface as opposed to dealing with raw simulation data. To that end, DCMS provides various user interfaces for data input and query processing.

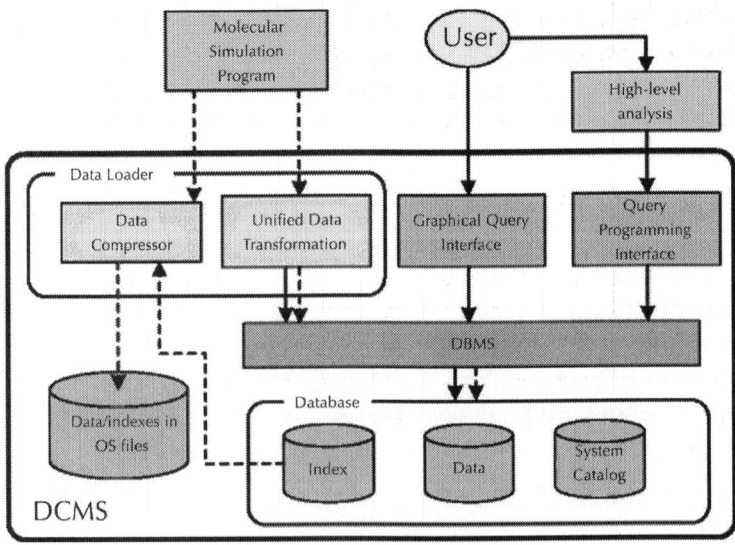

Figure 2: The DCMS architecture.

Data loader: The data loader is responsible for transforming simulation data to the format required for storage in the database system. First, it can read and understand data generated by current simulation programs and stored in popular MS file formats (e.g., GROMACS, PDB). We developed, as a part of the data loader, a *unified data transformation module* to perform such translations. A user only needs to specify the file format his/her data follows and data loading will be performed automatically. Second, the raw data will also be sent to a *Data Compression* module to generate a volume-reduced copy to be stored as compressed files. The rationale of this design is to enable efficient transmission of MS data among different data centers where DCMS are deployed. We will address this issue in Section "Data compression".

User interfaces: Runtime data access in DCMS is provided by a graphical query interface and an SQL-based query programming interface. In DCMS, we envision two types of user programs: *primary queries* and *high-level data analytics*. The primary queries correspond to (raw) data retrieval queries that can be directly supported by the DBMS via SQL. Analytical queries are those containing application-specific logic and are directly used by scientists for scientific discovery. The latter builds upon query results of the primary queries. To ease the development of high-level analytics, an important design of DCMS is that the query interfaces are extensible: an analytical query written by a user (called *user-programmed analytics*) can be integrated into the current DCMS query interface (and become part of the *DCMS built-in analytics* that can be directly accessed by an SQL statement). By this, the code of analytical queries can be reused by other users to issue the same query or build new analytical queries based on the current ones.

In addition to the query programming interface, all built-in analytics and primary queries can also be accessed from a graphical query interface, which accepts user query specifications via web forms and translates such specifications into SQL statements. The main purpose of designing a graphical interface is to, again, ease the use of DCMS to an extent that users can perform data analysis without writing programs.

MS data modeling and database design: Generally, an MS dataset contains a small number of physical features of a large number of atoms recorded at different steps of the simulation. Some of the important features include 3D locations, velocity, charge, and forces.

All data collected in one time interval is called a *frame*. The main body of an MS dataset is thus a collection of data items, each of which holds the information about an individual atom at a specific time frame. Conceptually, each data item can be viewed as a point in a multidimensional space with the dimensions representing the physical features we record. As a result, it fits the relational data model very well and we show the schemas corresponding to one simulation dataset in Figure 3. Note that an MS software such as GROMACS generates files holding information in rows. Naturally, our database design process started from a one-to-one mapping from those files to relations. We then refined the initial schemas by removing redundancy and reached a design shown in Figure 3.

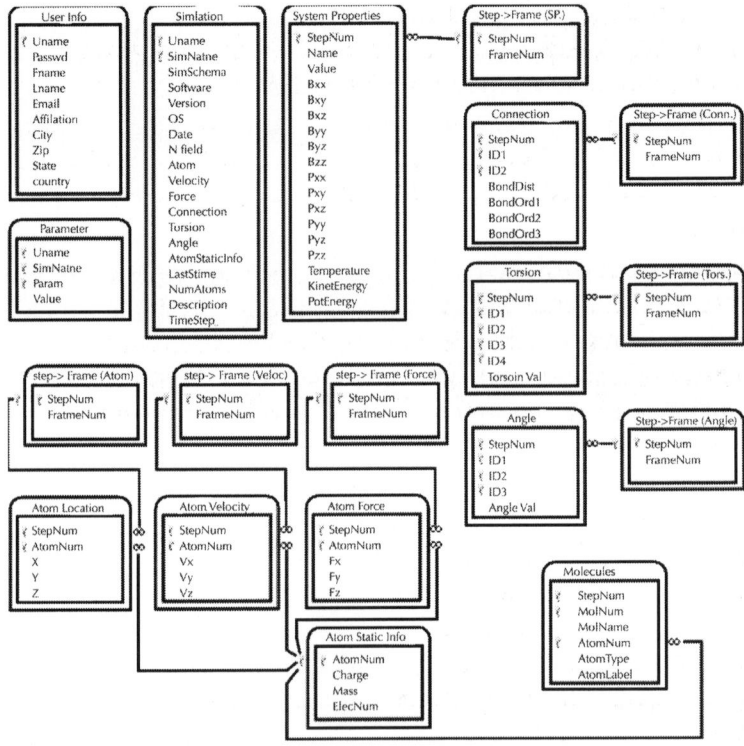

Figure 3: Schema of the DCMS database. The golden key symbol marks the primary keys and the lines represent foreign keys. Note that there exist foreign keys from all Ids in tables *Connection*, *Torsion*, and *Angle* to the *Atom Static Info* table. We did not draw them due to space limit.

First group of relations (i.e., *Simulation* and *Parameter*) describe time-invariant information of the simulation system. The *System Properties* table contains time-variant information of the entire system (instead of individual atoms). The *Atom Static Info* table holds the static features of an atom and forms a star-shaped schema with a series of other tables: *Location, Velocity, Force, Connection, Torsion, Angle,* and *Molecules.* The first three form the main body of the database - they represent atom states that change over time during the simulation. The reason why we cannot combine these three into one table is: MS programs usually do not output data at every step of the simulation, and different intervals (in steps) can be set to output location, velocity, and force. For the same reason, each of the three tables is linked to another table that maps the step number to the frame number. The next three tables hold information that describes atom to atom relationships. For example, a row in *Connection* represents a chemical bond between two atoms. Such relationships are time-variant therefore we again need to map their step numbers to frame numbers. The *Molecules* table is similar except it holds static relationship among the atoms. Specifically, each row in this table records the membership of one atom as a part of a molecule.

Analytical queries in DCMS: As mentioned earlier, DCMS provides system support for analytical queries that are unique in MS. The most popular DCMS built-in analytics are listed in Table 1, in which we assume an MS system has n atoms, and denote the mass, coordinates, charge, and number of electrons of an atom i as m_i, $\mathbf{r}_i$, q_i, and e_i, respectively. Roughly, such queries can be divided into two categories: (1) *one-body functions.* In computing this type of functions, the readings of each atom in the system is processed constant number of times, giving rise to $O(n)$ total running time. All queries in Table 1 except the last two fall into this category. Most such queries are defined within a single frame while the various autocorrelation functions are defined over two different frames; (2) *Multi-body functions.* The computation of these functions requires interactions of all atom pairs (2-body) or triplets (3-body). Popular examples include Radial Distribution Function (RDF) [16]-[18] and some quantities related to chemical shifts [19]. These functions are often computed as histograms. For example, RDF is generally derived from a histogram of all atom-to-atom distances (i.e., Spatial Distance Histogram (SDH)). Straightforward computation of multi-body functions is very expensive therefore it imposes great challenges to query processing.

Query Processing and Optimization in DCMS

In DCMS, we could rely on a legacy DBMS (e.g., Postgre SQL) to provide query processing mechanisms. However, we believe that explorations on the algorithmic issues in major DBMS modules (e.g., query processing and access methods) will further improve the efficiency of data analysis in DCMS. This is because existing DBMSs, with general-purpose data management as the main purpose, have little consideration of the *query types* and *user access patterns* that are unique in MS data analysis.

Primary Queries

Data indexing is the most important database technique to improve the efficiency of data retrieval. In DCMS, algorithms for processing primary queries will be exclusively index-based to reduce data access time. To support a rich set of queries, multiple indexes are necessary. However, it is infeasible to maintain excessive number of indexes due to the extremely high storage cost for MS data. Note that MS databases are most likely read-only therefore the maintenance (update) cost of indexes can be ignored. We have designed and tested several novel indexes to handle the various queries in DCMS but finally adopted the following indexes in our implemented system: (1) the B $^+$-tree and a bitmap-based index which are the default indexes provided by Postgre SQL - they provide a certain level of support for some of the MS queries; and (2) a new index named Time-Parameterized Spatial (TPS) tree to provide further performance boost. We accordingly modify the query optimizer of the DBMS to generate query execution plans that take advantage of the aforementioned indexes.

TPS Tree

For spatial (range, nearest neighbor, or spatial join) queries, Quadtree, R*-tree or SS-tree are the most popular indexes. The main challenge in DCMS comes from the continuous queries where time serves as an extra dimension therefore the above data structures are not suitable for MS queries. The design of TPS tree can be briefly described as *building a spatial index for each time frame in the dataset*. Then we need to combine neighboring trees to save space, taking advantage of the similarities among these trees. We decided to use Quadtree as the underlying spatial index to build TPS. This is because: (1) the performance of Quadtree in handling spatial queries is equivalent (sometimes even superior) to that of the R*-tree [20]; (2) the chances of getting an unbalanced tree (which is the main drawback of the Quadtree) are small due to the "spread-out" feature of MS data; and most importantly, (3) Quadtrees can be augmented to build other data structures needed for our high-level analytical query processing (Section "Analytical queries").

A major challenge in designing the TPS tree is to *minimize the storage cost*. Our main idea is to share nodes among neighboring Quadtrees corresponding to consecutive time frames - similar to the historical R-tree (HR-tree) [21]. The node sharing in an HR-tree depends on the assumption that some objects do not move for a period of time, which is not applicable to MS data where all atoms move at all times. However, we can exploit the existence of slow-moving atoms or those that move together to achieve node sharing. To build a TPS tree, we start by creating a spatial tree for each time frame using bulk loading; and then merge nodes in neighboring trees iteratively.

Analytical Queries

A straightforward view of analytical query processing consists of two stages: retrieve the raw data and compute the results in a separate program. While indexing techniques shown in Section "Primary queries" can speed up the first stage, further optimizations are made in DCMS by *pushing the computation into index construction*. To be more specific, we cache critical statistics of all atoms contained in a node of the TPS tree. Query results can be derived directly from such statistics, saving much time for visiting the raw data points. In the following text, we sketch our caching-based query processing strategy by visiting two groups of analytical queries.

One-Body Functions

Most one-body functions are *algebraic functions* [22] with features suitable for caching-based aggregation. Given a spatial region A in the simulated space, let us divide it into two disjoint subregions A_1, A_2. If we know relevant quantities for the computation a one-body function *f* of all atoms in both subregions, then the value of *f(A)* could be computed from those subregional quantities in constant time for most types of one-body functions. Take the center of mass (quantity*C* in Table 1) as an example: if we cache the values of the moment of inertia *I* and sum of mass *M*for both subregions, the mass center of region A can be computed by:

$$\frac{\sum_{i \in A} m_i \mathbf{r}_i}{\sum_{i \in A} m_i} = \frac{\sum_{i \in A_1} m_i \mathbf{r}_i + \sum_{i \in A_2} m_i \mathbf{r}_i}{\sum_{i \in A_1} m_i + \sum_{i \in A_2} m_i} = \frac{I_{A_1} + I_{A_2}}{M_{A_2} + M_{A_2}}$$

With the above reasoning, we can cache the chosen quantities in all nodes of a TPS tree without increasing the time complexity of tree construction. To compute the *C* value in an arbitrary region (Figure 4), we can traverse the tree and build *C* incrementally from the cached values of all nodes that are totally included in the query region. Thus, the running time depends on the number of tree nodes visited (instead of *n*). In fact, it is easy to see that the time complexity is $O(v^{2/3})$ where *v*is the volume of the query region. The savings on I/O time are also significant: cached quantities in a tree node can be orders of magnitude smaller (in size) than that of the raw MS data it covers [13].

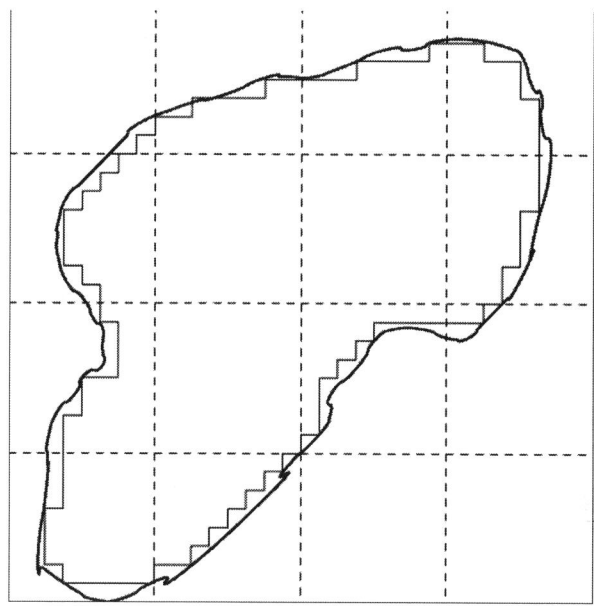

Figure 4: A 2D illustration of an irregular query region. Thin lines represent inclusive tree nodes visited at the 5th level in the Quadtree (the level 0 node in the tree covers the whole region).

Multi-Body Functions

The multi-body functions are all *holistic functions* [22] therefore cannot be computed in the same way as one-body functions. Current MS software adopts simple yet naïve algorithms to compute the multi-body functions [9]. For example, the SDH is computed by retrieving the locations of all atoms, computing all pairwise distances, and grouping them into the histogram - a $O(n^2)$ approach. For a large simulation system where n is big, this algorithm could be intolerably slow. In DCMS, we invested much efforts into algorithmic design related to such queries.

Our strategy for fast multi-body function computation follows the same path of the caching-based query processing. Specifically, we cache the *atom counts* in each spatial tree node and compute the multi-body functions based on these counts, avoiding the computation of pairwise interactions. Again, we do this on the TPS tree and call the atom counts of all nodes on one tree level a *density map* (which is basically the result of a *density function* query shown in Table 1). The main issue is how to translate the atom counts into a histogram. In the following text, we will sketch a series of algorithms we developed for SDH.

The main idea is to work on clusters of atoms (e.g., A, B, C, D shown in Figure 5) instead of individual atoms in deriving the function results. Beginning at a predetermined level, the TPS tree is traversed to see if any pair of nodes can form a distance range that is fully contained in a bucket of the histogram. This is the key operation in this algorithm and it is named *node resolving*. If a pair of nodes do not resolve, we keep resolving their child nodes in the tree. At leaf level, point-to-point distances between particles are computed to finish histogram processing. Such density-based algorithm achieves running time of $O(N^{5/3})$ [23].

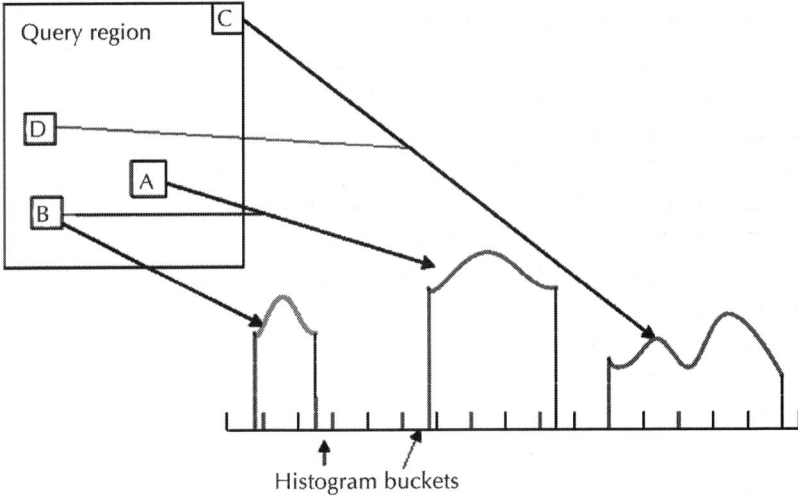

Figure 5: Intuition behind computation of SDH by considering pairs of nodes in a density function. Curves represent distribution of distances between two nodes (blue) or within a single node (red).

In many cases approximate query results are acceptable to users, therefore we developed approximate algorithms with much better performance to solve the SDH problem. Basically, we modified the node resolving operation to generate partial SDH faster. Whenever a desired error bound is reached while traversing the tree, the distance distribution into histogram buckets is approximated using certain heuristics. Such heuristics are constant time operations while having guaranteed error bound. Total running time of the algorithm is only related to a user-defined error tolerance. A detailed analysis of such algorithms is presented in [24]. Performance of SDH algorithms can be further improved if certain inherent properties of the simulation system are utilized. It is often observed in MS systems that the atoms are evenly spread out due to existence of chemical bonds and inter-particle forces. Therefore, it is possible to take advantage of the spatial uniformity of the atoms to speed up the computation of SDH. With the spatial distribution of atoms in nodes, we can derive the distribution of distances between any pair of nodes. The entire distribution (i.e., histogram) can thus be obtained by considering all such pairs. Similarly, the locality of atoms in different frames is also utilized to compute approximate SDH very efficiently [25].

Our algorithm performs the same operation on different pairs of regions. This gives us a hint to use parallel processing to further improve performance. General Purpose Computing on Graphics Processing Unit (GPGPU) is a low-cost high performance solution to parallel processing problems. Large number of parallel threads can be created on GPUs, which are executed on multiple cores. Unlike CPUs, the GPU architecture consists of more than one level of memory that can be addressed by the user program - the *Global memory* and the *Shared memory*. The latter is a cache grade memory with extremely low latency. This makes code optimization in GPUs a very challenging task. We developed and optimized GPU versions of above SDH algorithms and achieved dramatic speedup of computation [25].

View-based Query Processing

We developed a system-level approach to further improve the response time of analytical queries: *reuse the results of previous queries*. This is a generalization of the caching-based strategy and can be realized by defining a query as a *view* and caching its results (i.e., view materialization). In extreme cases where the same query is re-issued, it can be answered instantly by retrieving the recorded view. A more general situation is: new queries, although not identical to any materialized views, could benefit from one or more such views. For example, the results of the mass center (MC) queries against regions A, B, and C in Figure 6 are stored as views. To compute a newly-issued MC query against region Q1, we could utilize the MC of region A (from the corresponding view), with relevant values cached in TPS tree nodes corresponding to region Q1−A; to compute a query against Q2, we can first compute the results in region BUC (based on materialized views of B and C), and then process the query in region Q2− (BUC). Clearly, we save the time to compute query against regions A, B, and C from scratch in processing queries Q1 and Q2. As compared to the caching-based strategy, the view-based solution is more efficient since it is not constrained to visiting all tree nodes involved in the query region.

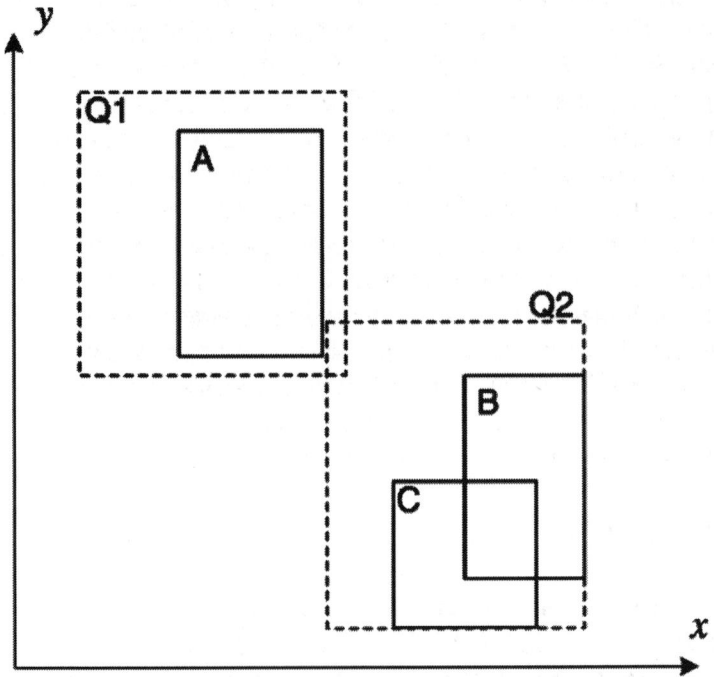

Figure 6: Two spatial queries (Q1, Q2) and three recorded views (A, B, C) in a 2D space.

To make the view-based solution work, the main challenge is the *design of query optimization algorithms that take views into account.*

Query optimizers of existing DBMSs are not established for our purpose: they focus on views that are built over various base tables [26], [27] in the database, often as a result of join operations. On the other hand, a view in our system maps a multidimensional data region to a complex aggregate. Such differences require development of novel techniques to address the following research problems.

View Representation

The first problem is *how to represent and store recorded views*. Since any view of interest in DCMS describes a certain type of query (e.g., mass center) over a collection of raw data points in a 3D region, we organize the views in data structures similar to the spatial trees used for indexing the data. We call this the *view index*, in which a leaf node has the form of $(BR, t_1, t_2, TYPE, rprt)$ where BR (short for Bounding Region, as in R-trees [28]) is a description of the query region of the view, $TYPE$ is a variable encoding the query type and parameters (e.g., resolution, maximum, and minimum of an SDH query), t_1 and t_2 refer to the starting and ending frame the query covers, and $rprt$ points to the query results of the view. An R*-tree-like structure is designed to organize the view entries based on their BRs. Upon receiving a query, we search the view index using the BR value of the new query as the key and retrieve all views that match the type and overlap with the query in their BRs and temporal coverage. The set of views retrieved form the basis for query optimization. We found the cost of maintaining and searching the view index is small due to its tree-based structure and the moderate number of views (as compared to the number of the data points).

Cost Evaluation of View-based Plans

Given a set of matched views, there could be multiple ways (i.e., execution plans) to execute a relevant query. For instance, to compute Q2 in Figure 6, we could either compute the result in region $B \cup C$ and merge it with that of $Q2 - (B \cup C)$, or merge the results of region $Q2 - C$ with that of C, or $Q2 - B$ with B. The query optimizer of DCMS should be able to list the different execution plans and choose one with the lowest expected cost. Obviously, those plans that do not involve any views should also be evaluated for comparison. For this purpose, *a cost model* for each query type is designed to quantify the time needed to accomplish a plan. Factors that are considered in the model include: area/volume of the relevant regions involved, expected number of data points in these regions, costs to resolve views with overlapping BRs (e.g., costs to compute the query in *BUC*, which can be used to solve Q2), and existence of indexes. For queries with a small number of execution plans, the decision on which plan to choose can be made by evaluating the costs of all the plans. We are in the process of designing heuristic algorithms to help make decisions with reasonable response time in facing a large number of execution plans.

View Selection

Storage space is the only cost in maintaining views in DCMS due to the nearly read-only feature of the database. However, this does not mean we can keep as many views as we want (even if enough storage space is available). The reason is that, when the number of views increases, the view index is packed with more and more entries with overlapping BRs. This has two undesirable effects: (1) since the search performance of R-tree-like data structures deteriorates when the nodes have larger overlapping regions, the time needed to retrieve relevant views from the tree increases; (2) the number of relevant views retrieved for a given query will also become larger. As a result, the number of execution plans increases exponentially. These make view-based query processing less attractive as the time used by the query optimizer in query evaluation could grow unboundedly. To that end, we need to discard materialized views from the view index and/or stop materializing new views when the performance gain of view-based query processing reaches zero. A scoring function with the following style is used to determine which views to discard/keep:

$$S = \frac{f}{o}$$

Where *f* is the frequency at which the view is utilized for query execution and *o* shows the extent to which the view's BR overlaps with other materialized views. In case of enforced view selection, a view with lower score will be discarded before one with a higher score. The intuition behind the above formula is to calculate the benefit to cost ratio of view maintenance: views that are frequently used for query processing (i.e., more benefit) and cover a less crowded region (i.e., lower maintenance/query optimization cost) should be kept. Note that storage cost is not included in the scoring function as we believe it is not the bottlenecking factor in our view-based query processing.

Data Compression

Simulation information is stored onto disk frame by frame for further analysis of the system under study. Given large number of particles in the system, a simulation of few micro-seconds can generate terabytes of data. The size of the data poses problems in input/output, storage, and network transfer. Therefore, compression of simulation data is very important. Traditional compression techniques can't achieve high compression ratio. Size of the data compressed using dictionary-based and statistical methods can still be large. Accessing a small portion of the compressed data requires decompression of entire data set. In addition, corruption of few bits can make the entire data unusable. Techniques that use spatial uniformity of the data, such as space-filling curves [29] can produce better compression ratios. But, all existing methods do not consider temporal locality for further compression. We group several frames together to form a window, and a window is compressed using our technique explained below.

In our framework (Figure 7), the data is first transformed, using principal component analysis (PCA), from 3D coordinate space to another 3D eigen space, with the dimensions sorted in decreasing importance levels in capturing the variance of the atoms' movements. In the eigen space, the discrete cosine transform (DCT) is applied to achieve lossy compression across a window of consecutive frames. One major design goal of our framework is that the lossy compression does not affect the results of the analytics that are often executed on MS data. Our technique fusing PCA and DCT enables our framework to: (1) perform balanced compression across all dimensions, (2) control and avoid compression errors and data corruption; and (3) allow more random access to any frame in the data – only a window of frames compressed together is accessed instead of fully decompressing the whole data file.

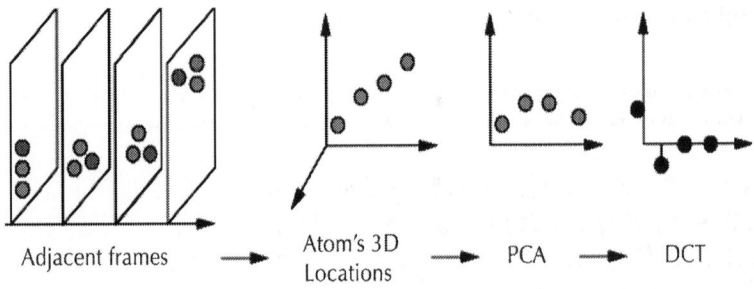

Adjacent frames ——→ Atom's 3D Locations ——→ PCA ——→ DCT

Figure 7: Steps in our MS data compression algorithm (borrowed from [[30]]).

Security Issues

Preserving data privacy is critical to organizations and research groups that employ external or third party analysts to understand the data, find out interesting patterns, or make new discoveries, but there are concerns on sharing the raw data. Sometimes, scientists have the same concerns over MS data. Privacy can be provided by database management systems through access control mechanisms (ACMs). ACMs limit the data access to users with special privileges, and ACM policies are directly supported by the SQL language. ACMs either restrict or completely grant access to the data. However, third party analysts may not be able to perform the best work without accessing other parts of the data that may depend on the private information. Attempts to provide flexibility ended with differential privacy mechanisms, which also face limitations due to requirements that are difficult to quantify, for both data providers and analysts. Therefore, the problem of preserving privacy from the range of ACMs to differential privacy is inadequately addressed. We performed some fundamental research on this topic within the context of DCMS. In particular, we designed an architecture named security automata model (SAM) to enforce privacy-preserving policies. SAM allows only aggregate queries, as long as privacy is preserved. Once it detects possible risk, differential privacy policy is enforced. It works on basic aggregate queries, liberating data owners from controlling special programs written by the analysts. Sequence of queries from all users in different sessions are monitored to detect the privacy breaches. We integrated this design into DCMS to address the question of how privacy can be defined, enforced, and integrated with existing ACMs using SAM.

DISCUSSION AND EVALUATION

We have implemented a prototype of DCMS and tested it with real MS datasets and workloads. We used Postgre SQL (version 8.2.6) as our database engine. We extended the current Postgre SQL codebase significantly by adding new data types, the TPS tree index, and various query processing and optimization algorithms. The TPS tree implementation, along with the addition of two new data types (i.e., 3D point and 3D box) was built on top of the SP-GiST [31] package. The query processing algorithms were programmed in following three ways: (1) most one-body functions are implemented as stored procedures using PL/pgSQL – a built-in procedural language provided by PostgreSQL; (2) the more complex multi-body functions are programmed as C functions and integrated into the PostgreSQL query interface; and (3) query processing algorithms related to the TPS tree are directly implemented as part of the PostgreSQL kernel. The data transformation module and data compressor were implemented as programs outside the DBMS. In the remainder of this 2, we report results of selected experiments run on our prototype. Such experiments were conducted on a Mac Xserve with quad-core Intel Xeon CPU running OS X 10.6 operating system (Snow Leopard), the server is connected to a storage array with a 35TB capacity with dual FibreChannel links.

Efficiency of Data Retrieval in DCMS

The main goal of this experiment is to show the efficiency of data retrieval in DCMS. Here we show the results of queries against a single MS dataset with 286,000 atoms and 100,000 frames, the total data size of which is about 250 GB. Note that such queries against a single simulation are typical in MS applications as comparing multiple simulations is less popular. This dataset was generated from our previous work to simulate a hydrated DPPC system in NaCl and KCl solutions [23]. For comparison, we sent the same queries against the data analysis toolkit of GROMACS – a mainstream file-based system for MS simulation and data analysis [9]. According to [6], GROMACS has better performance over other popular MS systems therefore represents the state-of-the-art in MS data analysis. We tested the following types of queries: (1) random point access (RDM); (2) single-atom trajectory retrieval (TRJ); (3) frame retrieval (FRM); (4) Range query (RNG); and (5) Nearest neighbor (NN) queries. These queries, as discussed in Section "Analytical queries in DCMS", form the basis for most high-level data processing tasks in MS. Upon loading the data, Postgre SQL automatically builds a B$^+$-tree index on the combination of step number and atom ID. Then we built the TPS tree on the *Atom Location* table for our experiments. For the control experiments using GROMACS, the queries were against the same dataset organized in 400 files, each holding 250 time frames. To achieve fair comparisons, we also used a grid-based spatial index in GROMACS [9] for the RNG and NN queries.

The main results of this experiment are shown in Table 2 where each number is the average of five experiments of the same query with randomly generated parameters. Clearly, the query performance is much better in DCMS, especially when we built the TPS tree. In fact, it achieves a speedup of 1-5 orders of magnitude over GROMACS. This shows the combined effects of record-based (rather than file-based) I/O and indexes. By the latter, we can directly visit the pages that hold relevant data records while we have to search through large files in the file-based solution. An interesting thing is that TRJ processing in DCMS, although faster than in GROMACS, still requires very long time when the TPS tree is not used. This shows that the existing indexes in Postgre SQL are not suitable for such queries. Apparently, all files are scanned in processing TRJ queries in GROMACS. Although indexes were used in GROMACS in processing the RNG and NN queries, we still observe a performance boost of 2-3 orders of magnitude in DCMS. The spatial index of GROMACS does not help at all in processing the RDM, TRJ, and FRM queries. In summary, the above results demonstrate significant improvement in data access speed by using DCMS for MS data processing.

Table 2: Query processing time (in seconds) in database-centric and file-based MS analysis

System	Queries				
	RDM	TRJ	FRM	RNG	NN
DCMS + TPS	0.0008	13.4	2.56	0.073*	0.029
DCMS	0.069	6239	2.48	0.122*	0.198
GROMACS	45.0	16410	52.5	8.49	16.8

*time depends on the query range

Kumar *et al.*

Kumar *et al.Journal of Big Data* 2014 **2**:9 doi: 10.1186/s40537-014-0009-5

SDH Computation Results

Synthetic data and real data generated from a collagen fiber simulation with 890,000 atoms are used for testing the performance of our SDH algorithms, with results summarized in Figure 8. Comparing with the brute force method, the density-based algorithm is about ten times faster under the larger bucket width. Once we consider the approximate algorithm, performance boost quickly reaches 2-3 orders of magnitude. Our ultimate solution that takes advantage of spatiotemporal locality of data further widens this gap - it is 3-4 orders of magnitude more efficient than the brute-force method. On the other hand, the errors generated from the approximate algorithms are very small (Figure 8(b)) - they are kept at the level of 1% for the density-based algorithm and go down to about 0.05% for the algorithm that considers spatiotemporal locality. Note that, the accuracy of our algorithms is not only small in practice, it is also guaranteed by a bound. Such details and other experimental results can be found in our previous publications [23], [24].

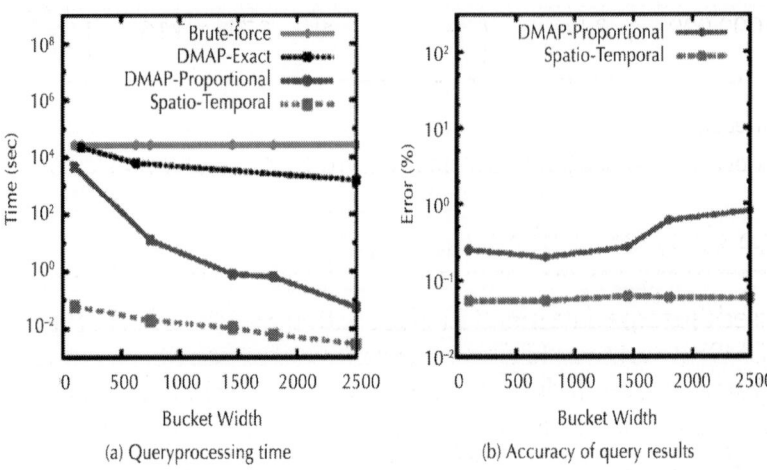

(a) Queryprocessing time   (b) Accuracy of query results

Figure 8: Performance of different SDH algorithms under different histogram resolutions.

We also implemented and experimented various algorithms based on GPUs. We developed our program under Nvidia's CUDA [32] parallel computing framework. The results of our implementation of the brute-force algorithm are shown in Table 3. Experiments were run on an Nvidia GTX570 GPU and a 2.66 GHz Intel Xeon CPU. We were able to achieve a speedup of up to 40 × against the CPU version while only using the Global Memory (GM) in the GPU algorithm. More efforts put into developing a Shared Memory (SM) based program yielded amazing returns of a 258 × improvement. The GPU version of the algorithm using spatial locality was also experimented, and reported a 20 × improvement (see [25] for more details). Based on our findings we believe that the GPUs are very powerful in processing complex analytics related to MS.

Table 3: Running time (seconds) of brute-force SDH method on GPU

System	GPU-time		CPU	GPU-speedup	
Size	SM	GM	Time	SM	GM
100,000	1.86	11.15	424.7	228	38
300,000	15.24	96.6	3812.2	250	39
800, 000	105.15	677.1	27142	258	40

8,000,000	3 hrs.	–	> 27 days	> 216	–

Kumar *et al.*

Kumar *et al.Journal of Big Data* 2014 **2**:9 doi: 10.1186/s40537-014-0009-5

Data Compression Results

We used the same data mentioned in Section "Efficiency of data retrieval in DCMS" to test the MS data compression method. Specifically, about 600 frames captured during the aforementioned simulation were used for our experiments. The following properties of atoms were stored in the file: x, y, and z coordinates, charge and mass measured. To measure the quality of compressed data, we use root mean square error (RMSE) to quantify the difference between the original data and the decompressed data. Figure 9(a) summarizes the main results. It can be seen that high compression ratio was achieved along with low RMSE between original and decompressed data. With very little loss of information, we achieved a compression ratio of at least 12 in such experiments. The effect of compression on the accuracy of analytical queries is also small (Figure9 (b)): for a shell resolution of 0.025Å, the difference between the RDFs of the original and decompressed data is negligible.

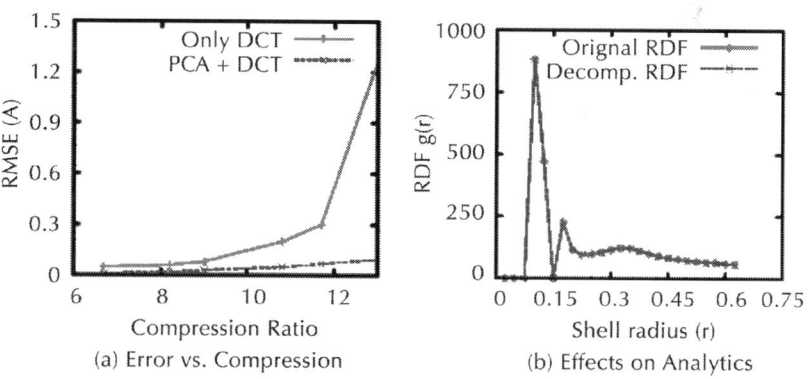

(a) Error vs. Compression   (b) Effects on Analytics

Figure 9: Performance of our MS data compression method: (a) compression ratio and error; (b) effects on radial distribution function (RDF). More details can be found in [30].

OTHER RELATED WORK

Traditionally, database systems are mainly designed for commercial data and applications. In recent years, the scientific community has also adopted database technology in processing scientific data. However, scientific data are different from commercial data in that: (1) the volume of scientific data can be orders of magnitude larger; (2) data are often multidimensional and continuous; and (3) queries against scientific data are more complex. The above differences bring significant challenges to system design in scientific databases.

In summary, scientific database research fall into the following three types. The first is to build databases on top of out-of-box DBMS products, as seen in the following examples: Gen Bank (http://www.ncbi.nlm.nih.gov/Genbank) provides public access to about 80 million gene sequences; the Sloan Digital Sky Survey [33] enables astronomers to explore millions of objects in the sky; and the Periscope [34] project explores declarative queries against biological sequence data. The second type focuses on extending the kernel functionalities of DBMSs to meet challenges in scientific data management. This includes work that deals with query language [35], data storage [36], [37], data compression [38], [39], index design [40], I/O scheduling [41], and data provenance [42]. The last type takes a more aggressive path by designing new DBMS architectures and building the DBMS from scratch. Most efforts along this direction happened in the past few years [43]-[46]. The SciDB project advocates a new data model (i.e., the multidimensional array model) for scientific domains and releases a prototype that enables parallel processing of data in a highly distributed environment. Clearly, our strategy of building DCMS falls into the second category.

CONCLUSIONS

Despite the importance of MS as a major research tool in many scientific and engineering fields, there is a lack of systems for effective data management. To that end, we developed a unified Information Integration and Informatics framework named Database-centric Molecular Simulations (DCMS). DCMS is designed to store simulation data collected from various sources, provide standard APIs for efficient data retrieval and analysis, and allow global data access to the research community. This framework is also a portal for registering well–accepted queries that in turn serve as building blocks for more complex high–level analytical programs. Users can develop these high–level programs into applications such as, applications that grant easy access of their data to experimentalists, visualize data, and provide feedback which can be used in the steering of MS. A fundamental component of the DCMS system is a relational database, which allows scientists to concentrate on developing high-level analytical tools using declarative query languages while passing the low-level details (e.g., primary query processing, data storage, basic access control) to DCMS. One of the most serious problems in existing MS systems is the low efficiency of data access and query processing. The unique query patterns of MS applications impose interesting challenges and also provide abundant optimization opportunities to DCMS design. To meet such challenges, we augmented an open-source DBMS with novel data structures and algorithms for efficient data retrieval and query processing. We focused on creative indexing and data organization techniques, query processing algorithms and optimization strategies. The DCMS system was also used as a platform to evaluate data compression algorithms specifically designed for MS data that can significantly reduce the size of the data.

Immediate work within DCMS will be focused on sharing computations (especially I/O operations) among different queries. Unlike traditional database applications, MS analysis normally centers on a small number of analytical queries (Table 1). Therefore, we can pro-actively run all relevant analytics at the time when the data is being loaded to DCMS. The advantage of this strategy is that only one I/O stream is needed - we have shown earlier that I/O can be the bottleneck in handling typical MS workloads. This requires us to modify the DBMS kernel to implement a master query processing algorithm to replace the ones dealing with individual queries. On the query processing side, utilization of other parallel hardware such as multi-core CPUs and FPGAs is definitely worth more efforts. Our current design of DCMS focuses on a single-node environment, deployment of DCMS on modern data processing platforms in a highly distributed environment (e.g., a computing cloud) will be an obvious direction for our future exploration.

AUTHORS' CONTRIBUTIONS

AK carried out the design, implementation, and experiments related to query processing, data compression, and data security. VG participated in experiments related to query processing and cache-based query optimization. MB designed and implemented index structures and the DCMS web interface. JF helped in the development of data loading and transformation module of DCMS. YT was in charge of the overall design of the DCMS system and algorithms related to analytical queries. XZ, SP, and YX participated in the system and web interface design. XZ also carried out the design and tuning of the data compression framework. YT drafted most parts of this manuscript while AK and VG also contributed to writing. All authors read and approved the final manuscript.

ACKNOWLEDGEMENTS

The project described here is supported by a research award (No. R01GM086707) from the US National Institutes of Health (NIH). Part of this work is also supported by two grants (IIS-1117699 and IIS-1253980) from US National Science Foundation (NSF), and gift from Nvidia via its CUDA Research Center program. The authors would like to thank the following collaborators for their contributions at various stages of this project: Anthony Casagrande, Shaoping Chen, Jin Huang, Jacob Israel, Dan Lin, Gang Shen, and Yongke Yuan.

REFERENCES

1.   Howe D, Costanzo M, Fey P, Gojobori T, Hannick L, Hide W, Hill DP, Kania R, Schaeffer M, St. Pierre S, Twigger S, White O, Rhe SY: Big data: The future of biocuration. *Nature* 2008, 455:47-50.

2.   Huberman B: Sociology of science: Big data deserve a bigger audience. *Nature* 2012, 482:308.

3.   Centola D: The spread of behavior in an online social network experiment. *Science* 2010, 329:1194-1197.

4.   Wu X, Zhu X, Wu G-Q, Ding W: Data mining with big data. *IEEE Trans Knowl Data Eng* 2014, 26(1):97-107.

5. J Bollen HM, Zeng X: Twitter mood predicts the stock market. *J Comput Sci* 2011, 2:1-8.

6. Michaud-Agrawal N, Denning E, Woolf T, Beckstein O: MDAnalysis: A Toolkit for the Analysis of Molecular Dynamics Simulations. *J Comput Chem* 2011, 32(10):2319-2327.

7. Humphrey W, Dalke A, Shulten K: VMD: visual molecular dynamics. *J Mol Graph* 1996, 14(1):33-38.

8. Nutanong S, Carey N, Ahmad Y, Szalay AS, Woolf TB: Adaptive exploration for large-scale protein analysis in the molecular dynamics database. In *Proceedings of 25th Intl. Conf. Scientific and Statistical Database Management. SSDBM*. ACM, New York, NY, USA; 2013:45-1454.

9. Hess B, Kutzner C, van der Spoel D, Lindahl E: GROMACS 4: Algorithms for Highly Efficient, Load-Balanced, and Scalable Molecular Simulation. *J Chem Theory Comput* 2008, 4(3):435-447.

10. Plimpton SJ: Fast parallel algorithms for short range molecular dynamics. *J Comput Phys* 1995, 117:1-19.

11. Brooks BR, Bruccoleri RE, Olafson BD, and States DJ, Karplus M: CHARMM: A program for macromolecular energy, minimization, and dynamics calculations. *J Comput Chem* 1985, 4:187-217.

12. Ng MH, Johnston S, Wu B, Murdock SE, Tai K, Fangohr H, Cox SJ, Essex JW, Sansom MSP, Jeffreys P: BioSimGrid: grid-enabled biomolecular simulation data storage and analysis. *Future Generation Comput Systs* 2006, 22(6):657-664.

13. Feig M, Abdullah M, Johnsson L, Pettitt BM: Large scale distributed data repository: design of a molecular dynamics trajectory database. *Future Generation Comput Syst* 1999, 16(1):101-110.

14. Finocchiaro G, Wang T, Hoffmann R, and Gonzalez A, Wade R: DSMM: a database of simulated molecular motions. *Nucleic Acids Res* 2003, 31(1):456-457.

15. van der Kamp M, Schaeffer R, Jonsson A, Scouras A, Simms A, Toofanny R, Benson N, Anderson P, Merkley E, Rysavy S, Bromley D, Beck D, Daggett V: Dynameomics: a comprehensive database of protein dynamics. *Structure* 2010, 18(4):423-435.

16. Frenkel D, Smit B (2002) Understanding molecular simulation: from algorithm to applications. Comput Sci Ser 1. Academic Press.

17. Bamdad M, Alavi S, Najafi B, Keshavarzi E: A new expression for radial distribution function and infinite shear modulus of lennard-jones fluids. *Chem Phys* 2006, 325:554-562.

18. Stark JL, Murtagh F: *Astronomical image and data analysis*. Springer, Berlin, Heidelberg; 2006.

19. Wishart DS, Nip AM: Protein chemical shift analysis: a practical guide. *Biochem Cell Biol* 1998, 76:153-163.

20. [www.cidrdb.org] Kim YJ, Patel JM (2007) Rethinking choices for multi-dimensional point indexing: making the case for the often ignored quadtree In: Proceedings of the 3rd Biennial Conference on Innovative Data Systems Resarch (CIDR), 281–291.

21. Nascimento M, Silva J (1998) towards historical R-trees In: Proceedings of ACM Symposium of Applied Computing (SAC), 235–240.

22. Szalay A, Gray J, vandenBerg J (2002) Petabyte scale data mining: dream or reality. Technical Report MSR-TR-2002-84, Microsoft Research.

23. Chen S, Tu Y-C, and Xia Y: Performance analysis of a dual-tree algorithm for computing spatial distance histograms.*VLDB Journal* 2011, 20(4):471-494.

24. Grupcev V, Yuan Y, Tu Y-C, Huang J, Chen S, Pandit S, Weng M: Approximate algorithms for computing spatial distance histograms with accuracy guarantees. *IEEE Trans Knowl Data Eng* 2013, 25(9):1982-1996.

25. Kumar A, Grupcev V, Yuan Y, Tu Y-C, Huang J: Computing spatial distance histograms for large scientific datasets on-the-fly. *IEEE Trans Knowl Data Eng* 2014, 26(10):2410-2424.

26. Halevy AY: Answering queries using views: A survey. *VLDB Journal* 2001, 10(4):270-294.

27. Afrati FN, Li C, and Ullman JD: Using views to generate efficient evaluation plans for queries. *J Comput Syst Sci* 2007, 73(5):703-724.

28. Guttman A: R-trees: a dynamic index structure for spatial searching. In *Proceedings of International Conference on Management of Data (SIGMOD)*. ACM Press, Boston, Massachusetts; 1984:47-57.

29. Omeltchenko A, Campbell TJ, Kalia RK, Liu X, Nakano A, Vashishta P: Scalable I/O of large-scale molecular dynamics simulations:

a data-compression algorithm. *Comput Phys Commun* 2000, 131:78-85.

30. Kumar A, Zhu X, Tu Y-C, Pandit S: Compression in molecular simulation datasets. In*4th International Conference on Intelligence Science and Big Data Engineering (IScIDE)*. Springer, Beijing, China; 2013:22-29.

31. Aref WG, Ilyas IF: SP-GiST: an extensible database index for supporting space partitioning trees. *J Intell Inform Syst* 2001, 17(2-3):215-240.

32. [http://www.nvidia.com/object/cuda_home_new.html] Nvidia. .

33. Szalay AS, Gray J, Thakar A, Kunszt PZ, Malik T, Raddick J, Stoughton C, vandenBerg J:The SDSS Skyserver: Public Access to the Sloan Digital Sky Server Data. In*Proceedings of International Conference on Management of Data (SIGMOD)*. ACM, Madison, Wisconsin; 2002:570-581.

34. Patel JM: The Role of Declarative Querying in Bioinformatics. *OMICS: J Integr Biol* 2003, 7(1):89-91.

35. Chiu D, Agrawal G: Enabling Ad Hoc Queries over Low-Level Scientific Data Sets. In*SSDBM*. Springer, New Orleans, LA, USA; 2009:218-236.

36. Arya M, Cody WF, Faloutsos C, Richardson J, Toya A: QBISM: Extending a DBMS to Support 3D Medical Images. In *ICDE*. IEEE, Houston, Texas, USA; 1994:314-325.

37. Ivanova M, Kersten ML, Nes N: Adaptive segmentation for scientific databases. In*ICDE*. IEEE, Cancún, México; 2008:1412-1414.

38. Shahabi C, Jahangiri M, Banaei-Kashani F: Proda: An end-to-end wavelet-based olap system for massive datasets. *IEEE Comput* 2008, 41(4):69-77.

39. Chakrabarti K, Garofalakis M, Rastogi R, Shim K: Approximate query processing using wavelets. *VLDB J* 2001, 10(2-3):199-223.

40. [www.cidrdb.org] Csabai I, Trencseni M, Dobos L, Jozsa P, Herczegh G, Purger N, Budavari T, Szalay AS (2007) Spatial indexing of large multidimensional databases In: Proceedings of the 3rd Biennial Conference on Innovative Data Systems Resarch (CIDR), 207–218. .

41. Ma X, Winslett M, Norris J, and Jiao X: Godiva: Lightweight data management for scientific visualization applications. In *ICDE*. IEEE Computer Society, Boston, MA, USA; 2004:732-744.

42. Chapman A, Jagadish HV, Ramanan P: Efficient provenance storage. In *SIGMOD Conference*. ACM, Vancouver, BC, Canada; 2008:993-1006.

43. [www.cidrdb.org] Stonebraker M, Becla J, Dewitt D, Lim K-T, Maier D, Ratzesberger O (2009) Requirements for Science Data Bases and SciDB In: CIDR 2009, Fourth Biennial Conference on Innovative Data Systems Research. .

44. [www.cidrdb.org]   Stonebraker M, Bear C, Cetintemel U, Cherniack M, Ge T, Hacham N, Harizopoulos S, Lifter J, Rogers J, Zdonik S (2007) One Size Fits All?- Part 2: Benchmarking Results In: CIDR 2007, Third Biennial Conference on Innovative Data Systems Research. .

45. Stonebraker M, Madden S, Abadi DJ, Harizopoulos S, Hachem N, Helland P: The End of an Architectural Era (It's Time for a Complete Rewrite). In *Proceedings of the 33rd* International *Conference on Very Large Data Bases*. ACM, University of Vienna, Austria; 2007:1150-1160.

46. [www.cidrdb.org] Sinha RR, Termehchy A, Mitra S, Winslett M (2007) Maitri Demonstration: Managing Large Scale Scientific Data (Demo) In: CIDR 2007, Third Biennial Conference on Innovative Data Systems Research, 219–224, Asilomar, CA, USA.

3

A Critical Review of Stall Control Techniques in Industrial Fans

Stefano Bianchi [1], Alessandro Corsini[1], Anthony G. Sheard[2], and Cecilia Tortora[1]

[1]Dipartimento di Ingegneria Meccanica e Aerospaziale, Sapienza Università di Roma, Via Eudossiana 18, Rome, Italy
[2]Fläkt Woods Limited, Axial Way, Colchester, Essex CO4 5ZD, UK

ABSTRACT

This paper reviews modelling and interpretation advances of industrial fan stall phenomena, related stall detection methods, and control technologies. Competing theories have helped engineers refine fan stability and control technology. With the development of these theories, three major issues have emerged. In this paper, we first consider the interplay between aerodynamic perturbations and instability inception. An understanding of the key physical phenomena that occurs with stall inception is critical to alleviate stall by design or through active or passive control methods. We then review the use of passive and active control strategies to improve fan stability. Whilst historically compressor design engineers have used passive control techniques, recent technologies have prompted them to install high-response stall detection and control systems that provide industrial fan designers with new insight into how they may detect and control stall. Finally, the paper reviews the methods and prospects for early stall detection to complement control systems with a warning capability. Engineers may use an effective real-time stall warning system to extend a fan's operating range by allowing it to operate safely at a reduced stall margin. This may also enable the fan to operate in service at a more efficient point on its characteristic.

INTRODUCTION

When a single fan operates in isolation the unstable aerodynamic condition, which we refer to as "stall" occurs at low flow rates. This type of stall varies according to fan type but is most severe in axial fans, forward-curved centrifugal fans, and backward-inclined centrifugal fans [1]. Fan stall occurs as the fan reaches its stable operating range limit. This happens when the pressure rise across a fan increases to the fan's pressure developing limit and the flow velocity though the fan reduces to the point at which it first falls to zero and then reverses. As the flow through a fan reverses, it separates from the fan blades with the turbulence that occurs with the separated flow buffeting the fan blades. This aerodynamic buffeting induces an increase in unsteady stress within the blades that can result in mechanical failure.

As a fan approaches stall, the separated flow initially occurs with one blade passage. Stall in one blade passage increases the aerodynamic blade loading on the adjacent blade passage, with a consequence that the "stall cell" moves to the next blade passage. This results in a cascading effect as a stall cell jumps from blade passage to blade passage. The shape of and distance between fan blades affect how the stall impacts fan performance with more highly aerodynamic loaded blade designs suffering a more severe reduction in performance during stall than lightly loaded designs. Centrifugal fans with radial blades show little change in performance in the event of stall. Radial-blade centrifugal fans do not rely on air passing through the fan and travel perpendicular to the centrifugal force which fan impeller rotation induces. As a result, stall is less of an issue in centrifugal fans generally than it is in axial fans.

Axial fans are particularly vulnerable to stall. Industrial fan manufacturers do not recommend axial fans for use in applications that require widely varying flow requirements unless a means of keeping flow rates above the stall point is available. Industrial fan manufacturers use proprietary antistall devices to control the flow in the axial fans' tip region. These anti-stall devices have the effect of stabilising the fan's performance. This eliminates the drop in fan performance at the point where it would have stalled without the anti-stall device, with the fan exhibiting a continuously rising pressure characteristic back to zero flow. This fan stabilisation is at the expense of fan efficiency, which typically reduces between 2 and 5 per cent with the presence of an anti-stall device. With an increasing focus on energy efficiency, anti-stall devices are becoming progressively less acceptable as industrial fan manufacturers strive to meet increasingly demanding minimum efficiency targets.

Historically, manufacturers have utilised anti-stall devices where a fan operates in conditions that may result in the fan stalling. However, application in which engineers do not expect the fan to stall can still result in stall. A fan can stall as a result of fan blade erosion or fouling or a significant increase in system pressure as a consequence of filters clogging. Additionally, a classical cause of industrial fans stalling is running them in parallel. When in parallel operation, one fan starting or stopping as others operate will inevitably result in the fan stalling during its starting and stopping transient. Consequently, poor fan maintenance, the blockage of filters within the system, or inappropriate control system programming can all result in fan stall.

A practice that engineers habitually employ in an attempt to avoid fan stall is oversizing of industrial fans for their application. System design engineers classically apply a safety factor to a fan's operating point when specifying industrial fans. Each engineer involved in the system's design adds his or her own safety factor. The result is that when finally installed, a fan operates on its characteristic far to the left of its optimum operating point. This lowers operating efficiency, with fans capable of achieving 80 per cent efficiency at their optimum operating point and frequently achieving less than 60 per cent when installed.

The European Union Regulation 327 became legally binding on January 1, 2013. This sets the minimum Fan and Motor Efficiency Grades (FMEGs) for industrial fans. The 2013 minimum fan and motor efficiency grades have resulted in approximately 33 per cent of fans sold before January 1, 2013, now being illegal within Europe as a consequence of not meeting the minimum fan and motor efficiency grade for their application. The European Union will raise minimum fan and motor efficiency grades on January 1, 2015. In the USA, the Department of Energy has been monitoring activity within the European Union. On February 1, 2013, the US federal government published a framework document in the Federal Register. This outlined the intended approach to fan regulation that aims to eliminate inefficient fans within the USA by 2019. The industrial fan community widely anticipates that the Department of Energy will adopt the same approach as the European Union, increasing the minimum allowable fan and motor efficiency within three years of introducing the initial 2019 targets. In practice, Asian countries take their lead on industry regulation from either Europe or the USA and with both now regulating, or declaring intent to do so; it is likely that Asian countries will do the same. Consequently, we may expect that over the next decade minimum fan or fan and motor efficiencies worldwide will first become mandatory, and second increase over time.

Given today's regulatory environment it is reasonable to assume that the pressure on design engineers will increase to develop fans with high peak efficiency. The practice of oversizing fans and fitting anti-stall devices will become progressively less acceptable as a consequence of the negative impact on fan efficiency. The result of specifying a fan such that it operates closer to its peak efficiency point when installed will increase the probability that the fan may stall. The peak efficiency operating point is invariably close to the fan's stability limits, and consequently erosion, fouling or filter clogging are more likely to result in stall. Therefore, the design of inherently "stall tolerant" fans, and development of a stall detection system for in-service use are becoming a rising priority for the industrial fan community.

Researchers have not systematically studied industrial fan in-service performance. However, advanced instrumentation and field testing can play a role in establishing where industrial fans actually operate on their characteristic and in those applications where a fan is prone to stall, assist in the verification and refinement of stall control techniques.

Thus, it is not enough to focus on the development of improved design-point performance. Researchers must develop a more complete picture of the challenges that occur with different industrial fan applications if they are to gain an insight into how they may improve fan efficiency without inadvertently producing fan designs more likely to stall in real world applications. This paper describes some of the competing perspectives on the physics that underpin fan aerodynamic stability and how knowledge of that physics can facilitate new industrial fan technology development.

STALL AND SURGE DYNAMICS IN AXIAL COMPRESSORS AND FANS

Predicting the conditions under which an aerodynamic instability will occur should be a standard part of the industrial fan design process. Over decades researchers have studied different forms of aerodynamic instability. Many studies have clarified the problem of axial flow compressor rotating stall, focusing on multistage machines [2–4]. Emmons et al.'s [5] earlier work was one of the first attempts to describe the mechanism underlying stall propagation. In general, two primary aerodynamic instabilities occur in decelerating rotors: (i) "rotating stall" in which regions of reversed flow occur locally; and (ii) "surge" in which periodic backflow over the entire annulus results in violent oscillations in the compression system [6]. Both forms of aerodynamic instability place mechanical stress on the rotors which can eventually lead to mechanical failure. Those researchers who have studied the subject report that strain gauge measurements on axial compressors indicate bending stress in blades exceed those measured during stable operation by a factor of five under rotating stall conditions [7]. Increasing bending stresses by a factor of five result in blade fatigue and consequently blade failure. The blade failure may occur whilst the fan operates under rotating stall conditions, but more usually does not. Typically, the fan's operation under rotating stall conditions results in initiating a fatigue crack. Once initiated, a fatigue crack may propagate under the influence of the bending stresses induced in the fan blades by normal operation. Consequently, a fan may fail due to fatigue days, weeks, or even months after operating under rotating stall conditions. By contrast, a surge can lead to the bending stress that increases to a magnitude at which mechanical failure occurs during the surge event itself.

Rotating stall is a progressive phenomenon, and, at least, initially does not necessarily result in the breakdown of a fan's pressure developing capability. Rotating stall, at least for axial machines [8], constitutes inception of the more severe flow instability, surge. Surge is a self-excited cyclic phenomenon, which affects the compression system as a whole. Large amplitude pressure rise and annulus averaged mass flow fluctuations characterise surge. It develops where a compressors constant speed pressure rise-volume flow characteristic line has an abrupt change in slope [8]. In Wo and Bons [9], the authors studied compressor performance, and reported experimental results that enabled them to conclude that a compressor's pressure rise-flow characteristic includes a region with positive slope. This indicates stall occurrence. Consequently, surge onset is dependent on both the compressor's characteristic and the system's characteristics into which it discharges.

Although we may regard rotating stall as a precursor to surge, the two constitute different aerodynamic phenomenon. The average flow during rotating stall is steady in time, but is circumferentially nonuniform. During a surge the flow is unsteady, but circumferentially uniform. It is as a consequence of a steady average flow with time that rotating stall may be localised within one or more of a compressor's stages. This has little or no effect on the system within which a manufacturer installs it. In contrast, the unsteady flow associated with surge impacts not only on the compressor, but the entire compression system.

Rotating stall and surge are distinctly different aerodynamic phenomenon but do share a common characteristic. We may regard both as the compression system's natural oscillatory modes [10–16]. Researchers are still debating whether rotating stall can result in centrifugal and single stage axial compressor mechanical failure, or if only surge can result in mechanical damage in these classes of rotating machines. The debate is inconclusive, and within the community that has studied rotating stall in centrifugal and single stage axial compressors there is disagreement as to the importance of rotating stall. This paper focuses on industrial fan technology rather than compressor technology, where researchers agree that rotating stall does result in mechanical damage and ultimately failure. Although a review of the effect of rotating stall in centrifugal and single stage axial compressors is beyond the scope of this paper, we review the extant literature within the context of its applicability to industrial fans. Our aim is to apply to the study of industrial fans the research from those academics and practitioners who focus on centrifugal and single stage axial compressors.

THE AERODYNAMICS OF STALL

For multi-stage axial compressors, rotating stall occurs at low shaft speeds and surge occurs at high speeds [8,17–27]. The distinction between low and high shaft speeds is a distinction between the ratio of pressure forces and flow momentum, which increase with increasing rotor speed. Recovering a multi-stage axial compressor from rotating stall is more difficult than recovery from surge [28]. Rotating stall is not a single phenomenon, but rather two distinctly different phenomena [29].

- Part span: where there is only a restricted blade passage region.
- Full span: the blade passage region is even smaller than in the case of part-span.
- Small scale: where a small part of the annular flow path is blocked.
- Large scale: where a large part of the annular flow path is blocked.

Surge has a more complex typology than rotating stall. We can distinguish at least four different surge categories with respect to flow and pressure fluctuations [8, 17, 30].

- Mild surge: a phenomenon associated with small pressure fluctuations and a periodicity governed by the Helmholtz resonance frequency. Flow reversal does not occur.
- Classic surge: a phenomenon associated with larger oscillations at a lower frequency than mild surge, also with no flow reversal. High frequency oscillations may also be present as the surge dynamics are nonlinear and introduce higher harmonics.
- Modified surge: a phenomenon associated with the entire annulus flow fluctuating in the axial direction, with rotating stall superimposed. This results in unsteady and non-axisymmetric flow. Modified surge is a mix of rotating stall and classic surge.

- Deep surge: a phenomenon associated with a more severe version of classic surge, where flow reversal occurs over the entire annulus.

When we consider an industrial fan's characteristics, we see that for a fixed blade angle and fan speed, as pressure across the fan reduces, flow increases. Bianchi et al. [31] studied the characteristics of an industrial fan, identifying the stable region over which reducing pressure results in increasing flow. In addition to the fan characteristic's stable region Bianchi et al. [31] characterised the fans unstable region, (Figure 1).

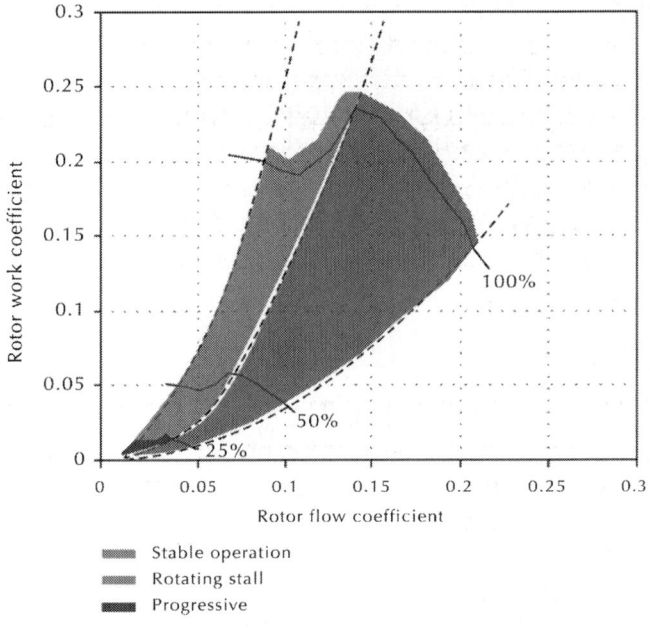

Figure 1: Regions of stable fan operation, rotating stall and progressive stall identified by Bianchi et al. [31] for an axial tunnel ventilation fan operating at 100 per cent, 50 per cent and 25 per cent fan design speed.

The resistance of an aerodynamic system increases with the square of the flow velocity through the system. Generating velocity, fan pressure increases with the square of velocity. If the required pressure is beyond the fan's peak pressure developing capability, the fan moves from the stable to unstable region. As a fan moves into the unstable region, both pressure and flow reduce. As the flow reduces, the required pressure to drive the flow through the system falls with the square root of speed. This results in the fan moving back into the stable region. As the fan operation stabilises, it generates additional flow and correspondingly, increasing system pressure until it drives into the unstable region again. This cyclic behaviour results in a hunting action that generates a characteristic sound similar to breathing.

An industrial fan's cyclic behaviour in surge may occur as a consequence of poor system design or leakage within the system. Classically, the systems into which engineers apply industrial fans include multiple branches with dampers fitted to enable flow to be directed down different branches at various times. If a branch in the system includes a damper that becomes stuck open, then this branch may result in the system becoming unbalanced, with a consequence that the fan may drive itself into an unstable region. In cases where the fan is operating primarily within the stable region with only occasional excursions into the unstable region, the fan can operate for extended periods of time without mechanical failure. In severe cases the fan motor will overload and overheat, and if the cyclic behaviour continues, fan blade mechanical failure will occur.

Stall Inception

The first challenge in attempting to identify appropriate approaches to stall control in industrial fans is to develop a fundamental understanding of the key physical phenomena which drive stall. The focus of any characterisation must be the stall inception process, as opposed to the characterisation of fully developed stall. Many researchers have characterised fully developed stall, with the research in the extant literature primarily focused on axial compressors. For a comprehensive review see Day and Cumpsty [3]. When considering the key physical phenomenon that drive stall in industrial fans, it is helpful to consider an industrial fans' tendency to exhibit cyclic behaviour as it moves from the stable to unstable region. A functional description of the processes at play during this cycle behaviour can provide the necessary insight to conceptualise, specify, and design a stall detection system.

Studying a subset of the published research scholars have conducted on industrial fans and in compressor research facilities facilitates the identification of key processes at play as an industrial fan moves from the stable to unstable region of its characteristic. Results that researchers have obtained in both types of rigs reproduce the physical phenomena at play within full scale compressors. A review of the results in low speed fan and scaled compressor facilities indicates that there is a hierarchy of possible stall inception mechanisms, starting with those that occur with low speed compressors and moving on to those that occur with multi-stage high speed compressors.

When we study the literature on low speed fan and scaled compressor facilities, it is apparent that two competing perspectives dominate the debate on stall inception and the physical mechanisms at play within rotating machinery. The first perspective focuses on long wavelength processes, or waves which span at least several blade pitches circumferentially. These waves constitute the primary physical process that determines compressor stability. The competing perspective focuses on short length scale events that are localised within one to four blade passages. Researchers consider these short length scale events as primarily responsible for stall inception. Although physical explanations of short length scale event significance dates back to Emmons, the concept that they may occur with stall inception is relatively new [5].

Several studies have suggested that some tip flow features in both compressors, low- and high-speed axial fans are directly responsible for generating short wavelength disturbances. The researchers studying short wavelength disturbances refer to them as "spikes" or "pips" that are responsible for localised part-span stall cells [32–35]. The stall cell's spike-like inception in a single stage is clearly evident in data that researchers obtained from a model fan [36]. We can see the spike-like inception at 28.5 seconds, Figure 2, with the flow then returning to its steady state condition for half a second before becoming unsteady at 29 seconds. Researchers studying stall inception mechanisms in industrial fans have correlated the spike-like inceptions with a change in fan acoustic emissions [37]. Other scholars studying the link between stall inception and acoustic emissions have utilised arrays of azimuthally distributed probes in an attempt to link the rotating unsteady pressure signals that they have measured in centrifugal pumps and compressors to their acoustic signatures [38, 39]. Kameier and Neise [40] and Bianchi and co-authors [31, 41] also studied the link between stall inception and acoustic emissions by establishing a link between tip-clearance noise and associated blade-tip flow instabilities in axial turbomachinery.

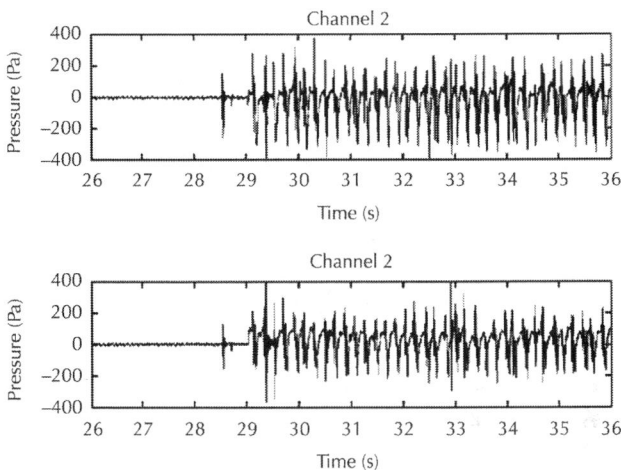

Figure 2: Data recorded from a pair of circumferentially offset high frequency response pressure transducers located over the blades of a 1250 mm diameter model of a variable pitch in motion induced draft fan. From 26 to 28.5, the fan is operating in the stable region of its characteristic. We can observe a spike-like feature at 28.5 seconds that is characteristic of stall inception. From 29 to 36 seconds the variation in pressure is associated with the fan's stalled operation [36].

Stall Development

There is a general consensus among researchers that stall is an instability phenomenon, local to the fan stage or rotor, in which a circumferentially uniform flow pattern ultimately results in completely blocking the annulus. As the fan blades become progressively more highly loaded, the stall commences with a "spike like" event and evolves into a rotating stall. The rotating stall classically evolves into a full stall or surge if there is a high enough system back pressure. A local stagnant flow region appears when the flow stalls. The regions propagate in the same direction as the blade rotation. This results in the stall region rotating around the annular flow path at a fraction of the rotor speed. The speed with which the stall rotates is typically between one-fifth and half of the rotor speed for fully developed stall. Initially, rotating stall cells rotate faster [29].

In reviewing rotating stall evolution, Cumpsty [37] noted that the drop in overall performance can occur as a so-called "progressive stall" or an "abrupt stall." Engineers usually associate the former with a part-span stall, which results in a small performance drop; whereas, they associate the latter with a full-span stall and a large drop in performance. Notably, the part-span rotating stall occurs typically in single blade rows [37] and usually leads to more complex disturbances in single-rotor or stage machines than in multi-stage compressors [4].

Mechanical Failure

Engineers have used strain gauge measurements on axial compressors [42] to measure bending stress in vanes that exceed stable operation by a factor of five under "rotating stall" conditions. Figure 3 illustrates an example of an industrial fan blade's mechanical failure that occurs with the unsteady mechanical loading that resulted from the fan stalling. In this example, stall resulted in a blade fatigue failure after operating approximately ten hours in a stalled condition. If this fan had been able to generate a back-pressure high enough to result in surge, the bending stress's heightening magnitude would have been enough to cause a mechanical failure during the surge event itself.

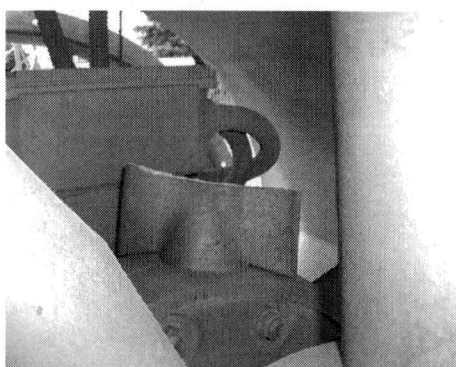

Figure 3: An example of a fan blade with a blade mechanical failure at the root aerofoil interface. This fan operated for approximately ten hours in a stalled condition before the mechanical failure [46].

An additional issue that industrial fan designers face is new legislation that governs the design of industrial fans that are intended for duel use. In this context, duel use refers to a fan use to both ventilate a tunnel or building during normal operation and clear smoke from escape routes in the event of a fire. Within the European Union it is a legal requirement to supply fans that are certified in accordance with EN 12101-3 requirements [43, 44], and outside the European Union the same requirements defined in EN 12101-3 are embodied within ISO 21927-3 [44, 45]. When extracting hot gas and smoke, an industrial fan's aluminium blades will grow thermally at a faster rate than the steel casing within which they rotate. Consequently, if the blades are not to touch the casing in the event of a fire the ambient blade tip-to-casing gap must be larger than would be the case if the fan were for ambient use only [42]. A consequence of increasing the blade tip-to-casing gap is typically a 20 per cent reduction in the fan's pressure developing capability. Fan designers frequently underestimate the impact of increasing the blade tip-to-casing gap on an industrial fan's pressure developing capability. A result of underestimating this reduction is that fans intended for dual-use operation are typically more prone to stall in service [46].

A particular feature of the environment within which industrial fans in tunnel ventilation applications operate is the pressure pulses that occur with trains moving through a tunnel. Pressure pulses can be up to ±50 per cent of the overall tunnel ventilation fan's work coefficient. Such pressure pulses drive the tunnel ventilation fan first up and then down its characteristic operating range [47]. To ensure that the tunnel ventilation fan continues to operate in an aerodynamically stable manner during this pressure transient, the tunnel ventilation system designer must incorporate sufficient margin to ensure that the tunnel ventilation fan does not stall due to the pressure pulses that occur with a train approaching and then moving away from a ventilation shaft.

A tunnel ventilation fan's propensity to stall under the influence of a pressure pulse is compounded when one operates at part speed. It is increasingly common to operate tunnel ventilation fans at part speed. Typically, the need for tunnel ventilation reduces at night, and therefore, one can achieve adequate cooling at a lower fan speed, and consequently at a lower operating cost. Although one may operate the tunnel ventilation fans at a lower speed, the speed of trains travelling within the tunnel remains constant, and therefore the pressure pulse magnitudes to which tunnel ventilation fans are subjected also remains constant. When a tunnel ventilation fan operates at 50 per cent speed, its pressure developing capability reduces by a factor of four. Consequently, a pressure pulse that could be accommodated at full speed will almost certainly drive the same tunnel ventilation fan into stall at 50 per cent speed.

As tunnel ventilation fan speed reduces, with a constant pressure pulse associated with trains passing the ventilation shaft within which the fan is located, there will be a critical speed at which a fan operating in supply mode stalls as the train approaches, or if the fan is operating in extract mode stalls as the train departs. Aerodynamic stall results in a significant increase in the unsteady forces applied to the fan blades. However, as the pressure pulse is transient the fan is not operated in a stalled condition for an extended period of time. Consequently the unsteady aerodynamic forces do not result in an immediate mechanical failure. However as a tunnel ventilation fan may be subjected to many pressure pulses each day, over time the cumulative effect of driving transiently into stall is to initiate a fatigue crack in one blade that then goes on to grow during stable operation until the blade mechanically fails.

We may conceptualise the impact of both positive and negative pressure pulses on a tunnel ventilation fan's operating point by referring to Figure 4. This provides an insight into how a fan adapts to a pressure pulse, with the duty point shifting up and down the fan characteristic under the influence of a +1000 Pa and −1000 Pa pressure pulse. It is custom and practice within the industrial fan community to assume that a pressure pulse may be modeled by shifting the system curve up and down by the magnitude of the pressure pulse. In Figure 4the + and −1000 Pa pressure pulse system curves are generated by shifting the system curve up and down 1000 Pa, respectively. Under the influence of a positive pressure pulse the fan operating point is assumed to shift from the fan duty point (black circle, Figure 4) to the point where the fan characteristic intersections the +1000 system curve (black square, Figure 4). Under the influence of a negative pressure pulse the fan operating point is assumed to shift from the fan duty point to the point where the fan characteristic intersects the −1000 system curve (black diamond, Figure 4). In doing so the fan is assumed to respond to a pressure pulse as if the change in pressure associated with the pressure pulse is slow in comparison to the reaction time of the fan.

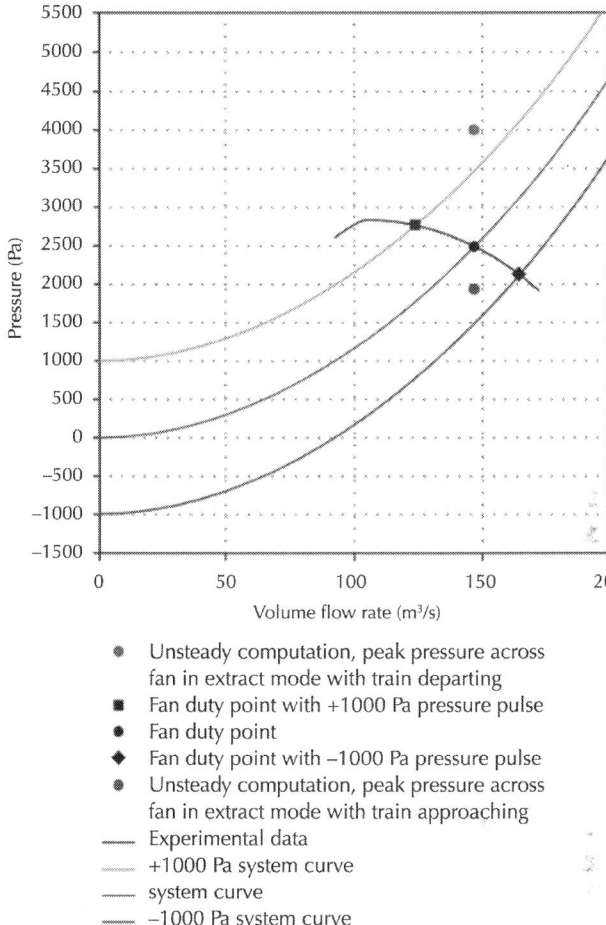

- Unsteady computation, peak pressure across fan in extract mode with train departing
- Fan duty point with +1000 Pa pressure pulse
- Fan duty point
- Fan duty point with −1000 Pa pressure pulse
- Unsteady computation, peak pressure across fan in extract mode with train approaching
- Experimental data
- +1000 Pa system curve
- system curve
- −1000 Pa system curve

Figure 4: The effect of a pressure pulse on a tunnel ventilation fan. Industrial fan designers have historically assumed that a fan runs up and down its characteristic in the presence of a pressure pulse (black symbols). Unsteady computational results for both a positive and negative pressure pulse indicate that the fan's operating point departs from its steady state characteristic during the transient associated with a pressure pulse (red and blue symbols). This departure results in unsteady aerodynamic forces increasing by a factor of two compared to those associated with operation of the fan at its duty point [47].

Recent research [47] suggests that tunnel ventilation fans do not respond to a pressure pulse as if the change in pressure associated with the pressure pulse is slow in comparison to the reaction time of the fan.

The interaction between changing pressure in a tunnel and the flow-field around fan blades is both transient and complex. When a pressure pulse within a tunnel is studied, it may be conceptualised as a change in volume flow rate though the fan. Unsteady computational results for a tunnel ventilation fan operating in extract mode [47] indicate that the impact of a train approaching the ventilation shaft within which a tunnel ventilation fan is situated is to unload the fan. The result is that the fan duty point shifts almost instantaneously to a lower pressure (blue circle, Figure 4). As a train passes the ventilation shaft a tunnel ventilation shaft with the tunnel ventilation fan operating in extract mode, the fan is over loaded. The result is that the fan operating point shifts almost instantaneously to a higher pressure (red circle, Figure 4). Significantly the time scales of this shift in operating point (from black circle to red circle and then back to black circle, Figure 4) are so rapid that the fan does not stall [47].

Despite the fan not stalling, unsteady forces on fan blades were shown to double in comparison to those associated with stable operation at the fans duty point [47]. This doubling of unsteady blade forces is significant. Within the industrial fan community designers generally believe that as long as a pressure pulse can be accommodated within the fan's pressure developing capability, as is the case in the example given in Figure 4, there is no mechanical consequence associated with pressure pulses. This is not the case and consequently, if blade designs are to avoid in-service mechanical failure, engineers must design them to accommodate the elevated aerodynamic forces that occur with pressure pulses.

STALL CONTROL TECHNIQUES

We typically derive flow control methodologies from an understanding of the relevant mechanisms or processes [48] and we can categorise them according to how one utilises flow control [49]. This can be

- active, entailing flow control; or
- passive, entailing a flow management.

Engineers have successfully applied passive and active stall control techniques into both industrial fan and compressor applications. However, passive stall control techniques are the norm in industrial fan applications and were the norm in compressor applications in the 1950s and 1960s. The drive to improve stability margin has been most intense within the compressor industry, and consequently, that is where the majority of active stall control research effort has taken place over the last two decades.

Active Control Systems

Active control systems monitor the event and its physics by relying on adequate warning or detection schemes in order to achieve the control objective. By contrast, passive control systems modify the flow dynamics in an effort to prevent the stall inception or to reduce the stall. Researchers have traditionally based the passive or preventive control concepts around blade or casing geometry modifications.

Active: Blade Pitch Control

Rotor pitch control is a technique that engineers mostly use in open rotors such as propellers or wind turbines to reduce the power when the air speed is above an allowable limit. With regards to axial fans, changing the angle formed by the blade's chord perpendicular to the axis direction constitutes a way to recover from stall. Lowering the pitch angle reduces the incidence angle onto the blade and reduces the blade loading. When we consider a variable pitch in motion (VPIM) fan's characteristic, it is apparent as the pitch angle reduces the fan's operating point migrates from the unstable to the stable region of the fans characteristic, Figure 5. It is the movement from the unstable to stable region that allows a reduction in blade angle to constitute a method by which a stall control system may recover a fan from stall. Bianchi et al. [36] studied experimental data from a variable pitch in motion fan as pitch angle reduced, observing that the pressure stabilised after 89 seconds, Figure 6, with reducing pitch angle. Consequently, a fan that was stalling with a 70° pitch angle no longer stalls when the pitch angle reduces to 50°.

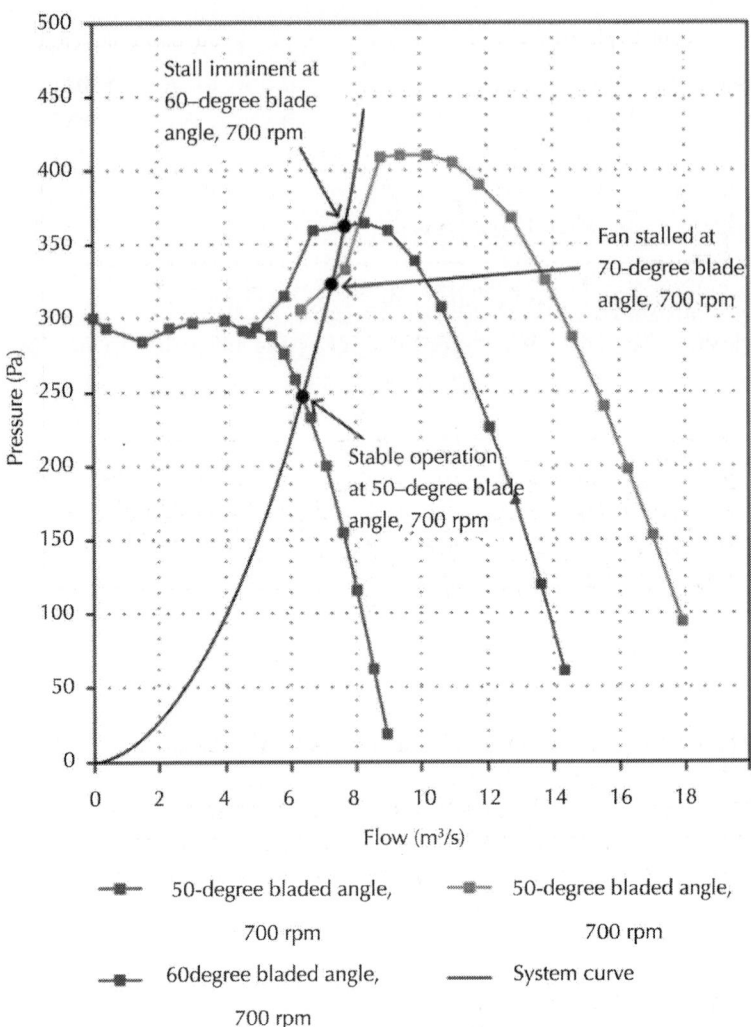

Figure 5: Stall recovery using variable pitch in motion (VPIM) blades. Operating points at 70°, 60°, and 50° pitch angle with all data taken at a rotational frequency of 700 rpm. At 70° the fan is operating in a stalled condition, to the left of the characteristics peak pressure. At 60° the fan remains stalled, with the fan operating just to the left of the peak in its characteristic. At 50° the fan is operating in the stable part of its characteristic, to the right of the characteristics peak pressure [36].

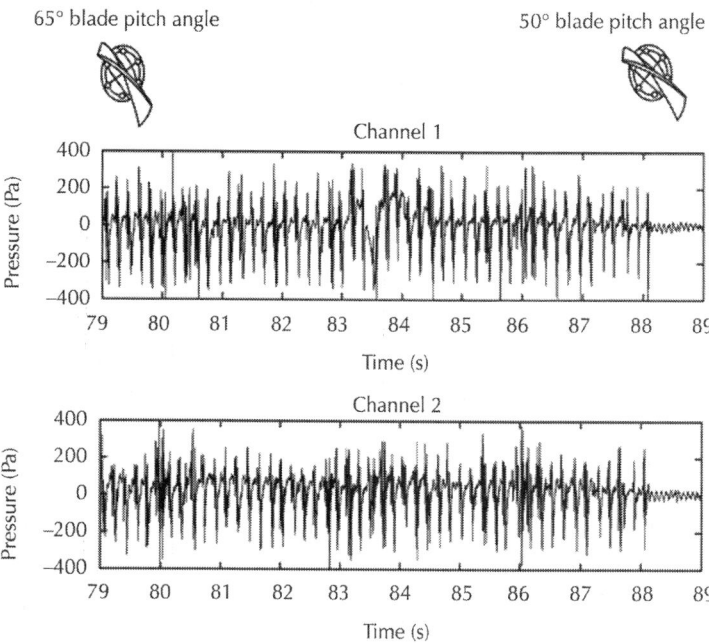

Figure 6: Data recorded from a pair of circumferentially offset high frequency response pressure transducers located over the blades of a 1250 mm diameter model of a variable pitch in motion induced draft fan. Blade angle reduces with increasing time, with the fan finally transitioning from stalled operation to stable operation at 89 seconds, at which time blade angle had reduced from an initial 70° at 79 seconds to 50° at 89 seconds [36].

Active: Rotational Frequency Control

A change in fan rotational speed does not result in a recovery from stall. Assuming that a fan is installed in a system with characteristics that obey the fan laws, a fan that is operating in the unstable region of its characteristic at 100 per cent speed will also be operating in the unstable region of its characteristic at reduced speeds. Therefore, reducing fan speed does not constitute a stall control method. Despite this reservation, reducing fan speed can protect a fan from the mechanical effects associated with operating in the unstable region of its characteristic. The direct mechanical stress in rotating components reduces with the square of speed. Consequently, reducing from 100 per cent to 50 per cent of design fan speed will reduce the direct mechanical stress in rotating components by a factor of four. However, operating a fan in the unstable region of its characteristic results in an increase in alternating stress induced in the fan blades as a consequence of the aerodynamic buffeting associated with stall.

Sheard and Corsini [7] studied the effect of operating a fan in the unstable region of its characteristic at full and part speed. They were faced with a particular problem with the fans supplied for an extension of the Athens Metro. Although the fans met their specification, during the summer when residents close to metro ventilation shaft portals were trying to sleep with their windows open, the noise from portals was loud enough to be problematic. The Athens Metro was asked to reduce night time portal noise emissions. A study of ventilation fan installations indicated that adding additional silencers was not practical. As there were multiple fans installed in each ventilation shaft, with only one required to run at night to supply the required flow of ventilation air, an option was to run multiple fans at reduced speed. Running a fan at reduced speed will reduce fan noise but reduce the flow of ventilation air. Running multiple fans will increase the volume of ventilation air, but as there are now multiple sound sources, the noise generated increases. Critically, the increase in noise associated with multiple sources will be less than the reduction associated with running multiple fans at reduced speed. Consequently, running multiple fans at part speed reduces overall portal noise emissions.

However, the reduction in fan pressure developing capability associated with part-speed operation was potentially problematic. The ventilation fans were subjected to a 500 Pa pressure pulse each time a train passed the ventilation shaft within which they were installed. When running at part-speed this pressure pulse would result in the ventilation fans driving from the fan characteristic's stable to unstable region each time a train passed the ventilation shaft. Although the ventilation fans would only drive into the unstable region of their characteristic transiently by a pressure pulse, with hundreds of trains passing each ventilation shaft each day, the cumulative effect would be the development of a fatigue crack in a blade followed by a mechanical failure.

Sheard and Corsini [7] studied the tunnel ventilation fan's mechanical performance. They fitted strain gauges to the blades, measuring the unsteady stress associated with operating the fan in both the fan characteristic's stable and unstable region at 100 per cent, 50 per cent, and 25 per cent design speed. In combination with the calculated direct stress in the fan blades at each speed, Sheard and Corsini [7] were able to derive a mechanical safety factor, Table 1. The results indicated that the tested fan may operate at 100 per cent speed in the stable region of its characteristic with a mechanical safety factor of 2.3. The same fan may also operate at 50 per cent speed in the unstable region of its characteristic with a mechanical safety factor of 2.5. As the safety factor 2.5 is larger than the safety factor 2.3, we may conclude that one may operate this fan at 50 per cent in the unstable region of its characteristic with less risk of mechanical failure than at 100 per cent speed in the characteristic's stable region. Consequently, reducing fan speed from 100 per cent to 50 per cent speed does not constitute a method of controlling stall, but does constitute a method of protecting the fan from mechanical failure in the event that operating in the fan characteristic's unstable part is unavoidable.

Table 1: Safety factors derived from strain gauge data for a fan at full and part speed [7]

Fan type		% design speed	Normal operating safety factor	Stalled operating safety factor
Plane stalling angle	casing, blade	100	2.3	0.3
Plane stalling angle	casing, blade	50	10.0	2.5
Plane stalling angle	casing, blade	25	106.0	7.3

Sheard and Corsini [7] extended their analysis, scaling the studied fans' characteristics from 100 per cent to 90 per cent speed. They observed that by doing so the fans' pressure developing capability reduced to a point where the 500 Pa pressure pulses to which this fan was subjected in practical application would take the fan to within 5 per cent of the fans' peak pressure developing capability. When Sheard and Corsini [7] scaled the studied fans' characteristics from 50 per cent to 55 per cent speed they concluded that a 500 Pa pressure pulse would still drive the fan from the stable to unstable region of its characteristic. However the increase in speed resulted in the mechanical safety factor reducing from 2.5 at 50 per cent design speed to 2.0 at 55 per cent design speed. From the above Sheard and Corsini [7] concluded that the tested fan could operate at up to 55 per cent design speed in the unstable region of its characteristic and down to 90 per cent design speed whilst remaining in the characteristic's stable region. The speed range between 55 and 90 per cent design speed was blocked in the fans' variable speed drive, and ensured that the fan only operated at speeds that did not put the fan at risk of mechanical failure under the influence of the pressure pulses.

Industrial fan manufacturers also utilise rotational frequency control to protect ventilation fans from the effect of unforeseen changes in system resistance. As the change in system resistance is unforeseen, it is not possible to predict the time when the change will occur. Bianchi et al. [31] studied a tunnel ventilation fan's stall characteristics using four unsteady pressure probes that they mounted on the fan casing whilst driving the studied fan into stall at 100 per cent, 50 per cent and 25 per cent of its nominal design speed. This allowed Bianchi et al. [31] to study the unsteady pressure signals that occurred with the fan's stable operation when instability was incipient and during stalled operation, Figure 7. Analysing the unsteady pressure signals enabled Bianchi et al. [31] to characterise the fan and identify the fan characteristic's stable and unstable regions at different fan speeds. From this, we may use an unsteady pressure measurement on the fan casing, in combination with the "blocked" speed range, 55 per cent to 90 per cent of design speed for the fan that Sheard and Corsini studied [7], as input for a control algorithm that establishes if a fan is mechanically at risk or may continue to operate without risk of mechanical failure.

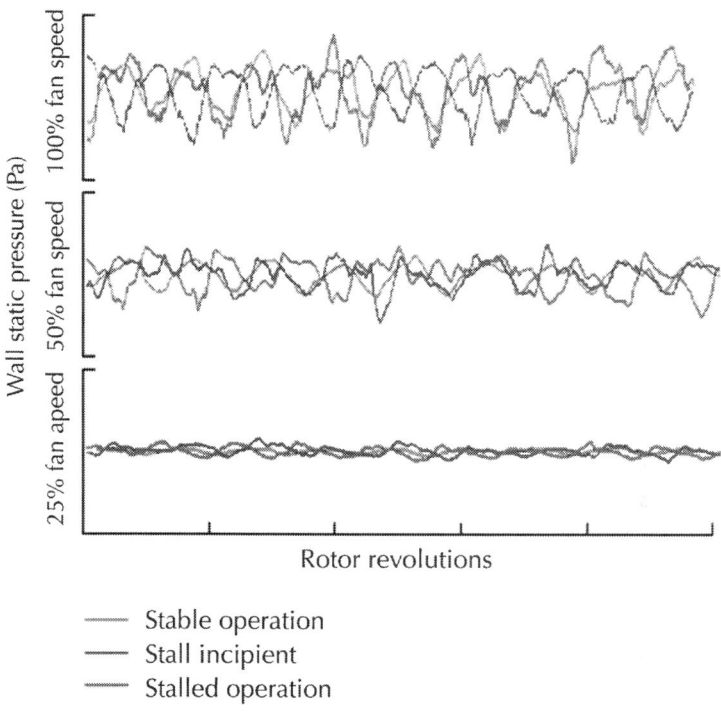

Figure 7: Low-pass filtered data (20 kHz) from a high frequency response pressure transducer located over the blades of a tunnel ventilation fan. The authors recorded the data during stable operation (green), when stall was incipient (blue) and during stalled operation (red) at full speed, (a) half-speed (b) and quarter-speed (c) [31].

Active: Air Injection

Researchers typically associate spike-like pressure pulses with stall inception in single rotor or stage industrial fans and compressors. A stall control technique that is effective in suppressing the spike-like pressure pulses is air injection. Air injection involves injecting high speed jets of air into the blade tip region that induces a transfer of momentum from the jet to the slower moving mainstream flow. The effectiveness of the high speed jets in suppressing the onset of stall is linked to the jets' influence on the tip clearance vortex's evolution and other flow features that occur with the over-tip blade flow.

Researchers have extensively studied air injection. Suder et al. [50] proposed a discrete tip injection technique, and Nie et al. [51] and Lin et al. [52] based their proposal on microair injection. Whilst requiring significant power to drive the associated control system actuation, these control techniques result in 5 to 10 per cent improvement in compressor stall margin. More recently, researchers have studied the underlying flow physics that occur with flow-field excitation in the blade tip-to-casing region using a spatially distributed actuation system to control the blade tip leakage vortex's evolution [53]. The researchers' hypothesis is that controlling the blade tip leakage vortex's evolution will promote the tip vortices' dissipation and therefore will suppress a part of the flow structure involved in spike formation.

A potential advantage of a spatially distributed actuation system to control the blade tip leakage vortex evolution is the low power requirement that occurs with control system actuators. The necessary power requirement to drive a control system's actuators constitutes a loss of efficiency for the industrial fan or compressor to which the control system is fitted. In an effort to minimise the power requirement, Vo et al. [54] proposed the use of acoustic actuation and Corke and Post [55] proposed magnetic actuators. Interest remains high in air injection within the compressor community, with Vo [56] proposing a method to suppress rotating stall inception in multi-stage compressors around the compressor's full circumference, Figure 8. At the time of writing the use of air injection is limited to compressor applications as a consequence of the complexity and cost of the technology. The industrial fan community continues to monitor the compressor community's progress, but currently there is no active research aimed at transferring this technology to industrial fans.

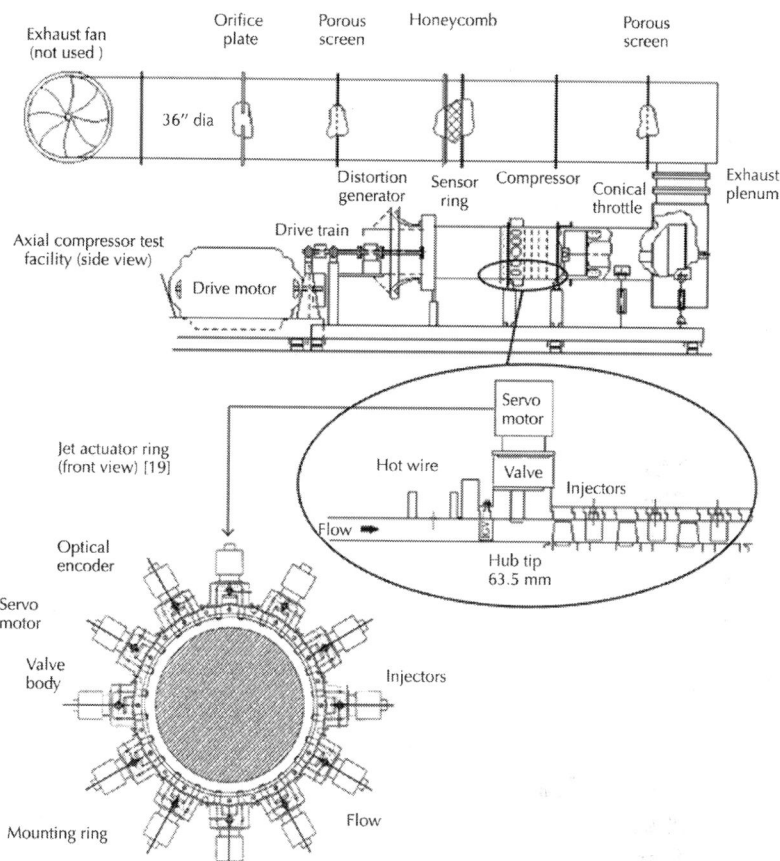

Figure 8: The Massachusetts Institute of Technology (MIT) low-speed three-stage axial compressor test facility with jet actuation [56].

Active: Bleed Valves

Researchers have tested stall control with bleed valve actuation on a small scale with compressors at low speed [57, 58]. The bleed valve opens to suppress the onset of stall. The technique aims to maintain the average flow through a compressor above the compressor's critical flow, below which the compressor blades will stall. The operator maintains average flow using bleed air from the compressor discharge that increases the flow through the compressor's lower pressure stages.

Prasad et al. [57] have presented two schemes for using bleed valve actuation: bleed air back into the compressor inlet and bleed air back into a recirculation plenum inlet. In the former, the bleed air does not affect the flow through the inlet; however, the bleed does decrease the flow to the combustor. In the latter, the flow into the combustor is equal to the inlet flow with the re-circulated air reducing the compressor load. The second scheme is more effective as bleed recirculation delays the onset of rotating stall and the delay increases with the magnitude of recirculation. Recirculation reduces flow into the combustor by altering the compressor's operating point, and consequently, an operator should not use recirculation continuously, but only when stall is imminent.

Passive Control Systems

Researchers base passive approaches to stall control on techniques that modify the flow field in the blade tip-to-casing region. Passive approaches date back to the 1950s, when researchers first utilised casing treatments in axial compressors. Skewed slots and grooves cut into the casing above the rotor improved stall-margin, with grooves both improving stall-margin and having the lowest impact on compressor efficiency [59]. The relatively low cost of passive approaches has resulted in industrial fan manufacturers historically favouring them. Consequently, whilst the compressor community today utilises primarily active control approaches, industrial fan manufacturers are still developing and refining passive approaches.

Passive: Stabilisation Rings

Industrial fan manufacturers have historically favoured the stabilisation ring, fitted to the fan casing, as the preferred antistall device. As an axial fan approaches stall, the flow velocity through the fan reduces and the axial fan blades increasingly act as a centrifugal fan impeller. Although antistall devices have evolved, their most common present day embodiment consists of a stabilisation ring placed around a fan casing immediately upstream of the fan blades' leading edge, Figure 9. As an axial fan approaches stall, the flow velocity through the fan reduces and the flow progressively centrifuges towards the blade tips.

At a critical pressure across the fan, the flow velocity falls to zero, and the flow in the blade tip region reverses. The stabilisation ring is able to stabilise fan performance as it contains a set of static vanes. These static vanes redirect the reverse flow into an axial direction, and then reintroduce it into the mainstream flow upstream of the blades, Figure 10. This stabilises the fans characteristic, with the fans now exhibiting a pressure characteristic that rises continuously back to zero flow.

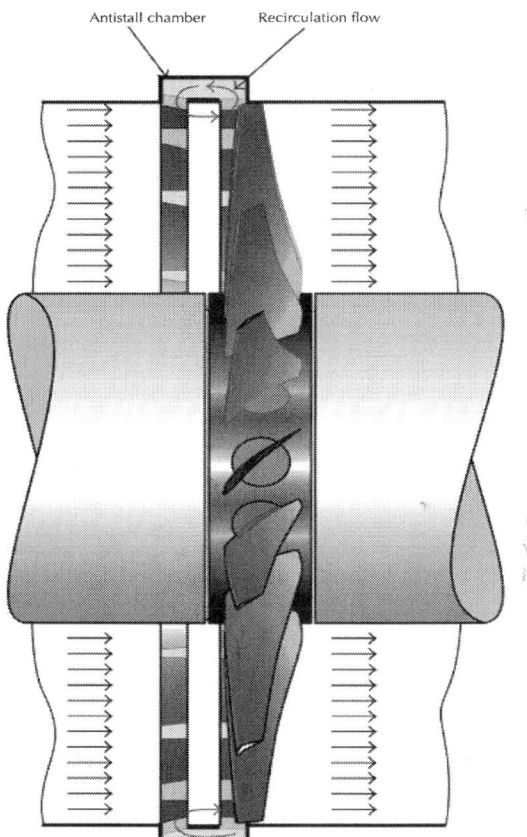

Figure 9: An axial flow fan fitted with an anti-stall ring. The anti-stall ring comprises an extension to the fan casing just over and upstream of the blades. The anti-stall casing incorporates static vanes, shown in yellow. As a fan approaches stall the flow though the fan is centrifuged up the blades and stalls as the flow spills out of the fan inlet. The blades redirect the flow in an axial direction, and reintroduce it upstream. This process of straightening and re-

introducing the flow stabilises the fan's performance, eliminating the drop in pressure developing capability classically associated with a fan operating in a stalled condition [72].

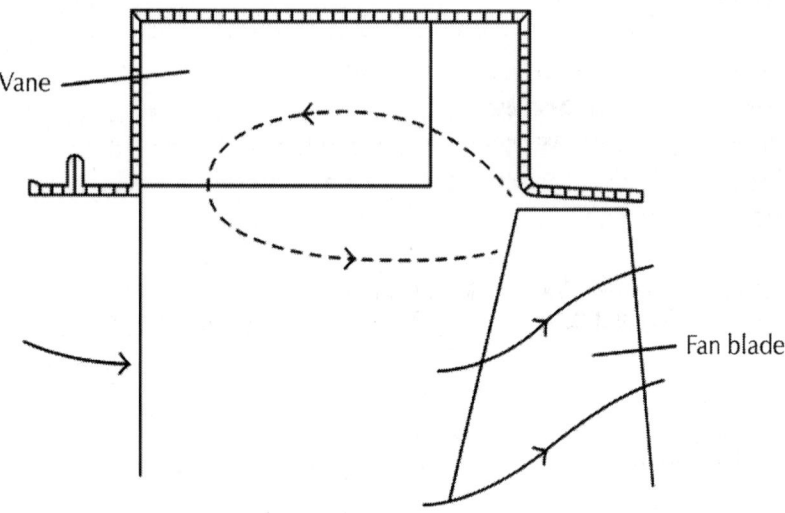

Figure 10: The proposed "stabilisation ring" arrangement adapted from Karlsson and Holmkvist [61] by Bard [73].

In 1965, Ivanov patented the first stabilisation ring [60]; however, the use of a full set of guide vanes upstream of the fan was difficult to apply in practical applications. Later Karlsson and Holmkvist [61] developed the stabilisation ring concept that incorporated the static vanes into a ring fitted around the fan casing. Despite the effectiveness of Karlsson and Holmkvist's stabilisation ring concept, it does have one unintended negative consequence. A fan fitted with a stabilisation ring will lose between 2 and 5 per cent efficiency as a direct consequence of the stabilisation ring [46].

The advent of the Energy using Product (EuP) Directive within the European Union has resulted in a mandatory minimum Fan and Motor Efficiency Grades (FMEGs) that became legally binding on 1 January, 2013. The minimum FMEGs will increase on 1 January, 2015. The industrial fan community widely expects that at some point in the future minimum allowable FMEGs will effectively render stabilisation rings obsolete as a consequence of their negative effect on fan efficiency. At some point in the future the reduction in efficiency that engineers associate with stabilisation ring applications will result in the fan's FMEG falling below the allowed minimum.

Houghton and Day [62] present a possible way forward for industrial fan designers who are no longer able to utilise stabilisation rings, demonstrating that a compressor's stall resistance may be improved by incorporating a groove into the compressor's casing. The groove was located approximately 50 per cent of blade chord upstream of the blade's leading edge. Incorporating a groove into the compressor casing did not result in reducing compressor efficiency, and consequently, a possible avenue of endeavor for industrial fan designers who are no longer able to apply stabilisation rings because of their negative effect on fan efficiency is to study casing groove application into industrial fan casings.

Passive: Air Separators

Air separators are able to effectively suppress the onset of stall. Yamaguchi et al. [63] designed an air separator which has radial vanes with their leading-edges facing the fan rotor blade tips so as to "scoop" the tip flow, Figure 11. The air separator differs from a stabilisation ring in that air separator vanes are radial, in contrast to a stabilisation ring vanes that are axial. Yamaguchi et al. [63] studied the air separator and analysed its stall suppression effects on a low-speed single-stage, lightly loaded axial flow fan. In the air separator's recirculation passage downstream from the inlet cavity, a series of circumferential vanes correct the swirl flow in the axial direction. When the fan approaches stall there is an increase in swirl speed and centrifugal force on stall cells. This causes the stall cells to centrifuge spontaneously into the air separator's inlet. Air separators therefore separate stall cells from the main flow, and as a consequence of requiring no moving parts, constitute a passive stall control method.

Figure 11: An axial fan mounted in a casing containing an air separator [63].

The four active and two passive stall control techniques, Table 2, each represent a valid approach to the control of stall. Researchers have developed and utilised each in different industrial fan or compressor applications where the technique has proven effective. In practice it is stabilisation rings that are the most widespread application in industrial fans and bleed valves that are the most widespread application in compressors. The other active and passive stall control techniques are the subject of research in an on-going effort to better understand the flow physics that underpin stall and development in order to improve industrial fan and compressor stability.

Table 2: Matrix of stall control techniques

Technique	Classification	How it works	Results
Blade pitch control	Active	Changes the fluid dynamics; changing the pitch angle.	Acts after stall detection [36].
Rotational frequency control	Active	Changes the fluid dynamics; controlling the rotor speed.	Acts after stall detection [31].
Air injection	Active	Rotating stall inception is achieved using full-span distributed jet actuation.	The nonideal injection reduces the stall range's extension by about the same proportion of the effective pressure increase [56].
Bleed valve	Active	Maintain the average flow through the compressor above the critical flow. The air bleeds from the plenum, so as to increase the flow through the compressor.	Recirculation alters the compressor operating point. Only use the recirculation or ambient bleeding when stall is imminent [57, 58].
Stabilisation rings	Passive	Changes the fluid dynamics; provides the stalled flow with a route back into the impeller.	Not sure that the stall occurs [72].
Air separator	Passive	A series of circumferential vanes correct the swirl flow in the axial direction; the stall cells centrifuge spontaneously into the air separator inlet and separate from the main flow.	Air separators are able to effectively suppress the stall zone after the stall cells appear [63].

STALL DETECTION SYSTEMS

Stall control techniques have proven effective in service; however, they are inevitably reactive. Stall control techniques require the fan to be stalling before they have any effect. In many applications it would be more appropriate for the control system to take action to avoid stall before it occurs, as opposed to managing the consequences of stall having occurred. For a control system to take action to avoid stall it is necessary to first predict stall onset. Predicting stall onset is challenging, and is an ongoing research subject in both the industrial fan and compressor community. Despite the challenges of predicting stall onset, it remains an essential precursor to the development of a more effective stall control system.

Stall detection systems that identify stall onset have the potential to form an input into a proactive stall control system theoretically capable of reacting before a fan actually stalls. Researchers recognise that studying stall detection both in industrial fans and compressors as critical to developing a stall management system. To form an effective input into a stall management system, a stall detection system requires as input the output from high frequency response sensors located in the industrial fan or compressor blades' immediate vicinity. Researchers then use the output from high frequency response sensors to identify stall precursors, and when detected, to generate a warning signal that inputs into the stall control system. The stall control system is then able to take remedial action to prevent the identified stall precursors from developing further.

Wadia et al. [64] and Christensen et al. [65] have proposed stall management systems based on instantaneous near-field pressure measurements. They have studied the effectiveness of stall detection systems as part of a stall management system when applied to a multistage high-speed compressor test rig. A challenge that both research teams faced was the very short time between the identification of stall precursors and compressor stall. Although stall detection systems have the potential to provide a useful input into compressor stall management systems, those stall management systems must be capable of reacting within a few milliseconds if they are to use an input from a stall detection system effectively.

Two-Point Spatial Correlation

In axial fans instabilities occur primarily as wave-like disturbances around the annulus in the circumferential direction. In the initial state of instability the disturbance amplitude is small, but increases with the evolution of the instability. One may use spatially adjacent fast-response pressure transducers, microphones or hot wire anemometers to identify stall precursors. Researchers associate these precursors with the formation of three-dimensional disturbances of finite amplitude, located in the blade tip region. These are characterised by a spike in the signal which fast response transducers record.

In order to accurately identify the spike in a signal recorded by a fast response transducer as a stall precursor, it is important to characterise the stall inception process dynamics. Only in characterising the stall inception process can one distinguish between spikes that occur with stall inception from background noise. Spike isolation is possible using a windowed two-point spatial correlation which provides spatial and temporal information about rotating features in the flow [66]. The windowed two-point spatial correlation technique is insensitive to low pass filtering and parameter selection over a wide range of values and is valuable for analysing both pre-stall and stall inception behaviour [66].

Stochastic Model

The stochastic model for detecting stall precursors utilises an autocorrelation technique. The signals from two circumferentially off-set high frequency response pressure probes mounted in the industrial fan or compressor casing close to the blade tips are autocorrelated. The correlation typically decreases as the compressor or fan approaches its stability limit, and therefore tracking the correlation provides a measure of the industrial fan or compressor's proximity to its stability limit. Dhingra et al. [67] developed the stochastic model. They developed an autocorrelating algorithm and established a minimum threshold correlation value that corresponded to the imminent onset of compressor stall. Although able to demonstrate that the stochastic model could form the basis of a stall detection system, the researchers conducted their reported work in a laboratory and did not include the development of a stall management system that utilised the stochastic model output as an input.

Travelling Wave Energy Analysis

An alternative modelling approach to either the two-point special correlation or stochastic model is travelling wave analysis. Travelling wave analysis involves calculating wave energy, which we define as the difference between positive and negative frequency power spectra. We then compute an "energy index" for a fixed-time window that must extend to include the spatial Fourier modes within it. Tryfonidis et al. [68] developed the travelling wave analysis as a real time measure of compressor stability. By providing a real timer measure of compressor stability, the analysis technique is useful for providing early warning of spike-type stall inception in high-speed compressors.

Cross-Correlation Analysis

Cross-correlation analysis is a further analytical approach that cross-correlates a pair of near-field pressure signals. Developed by Park [69], the analytical approach is based on the observation that short wavelength disturbances that are recognisably spike-like indications of incipient stall form and decay many rotor revolutions before stall occurs. By cross-correlating the signal from a single high frequency response pressure sensor, we can correlate the presence of spike-like pulses in the pressure signal from one rotor revolution to the next. The analysis produces a similar output to that of the two-point spatial correlation technique but requires only one sensor and is therefore more practical in real world stall detection system embodiments.

Acoustic Stall Detection

The symmetrised dot pattern (SDP) stall detection technique is based on an industrial fan or compressor acoustic pressure signal's visual waveform analysis [41]. The symmetrised dot pattern technique differentiates between critical and noncritical stall conditions and provides a form of visualisation that enables one to identify stall precursors. Bianchi et al. [36] first reconstructed the symmetrised dot patterns from the unsteady pressure which they measured on an industrial fan's casing and more recently [70] demonstrated the proposed technique's validity using a symmetrised dot pattern technique from sound pressure signals. This enabled them to extend the technique to the use of measured acoustic pressure signals in different locations in an industrial fan's acoustic far field.

The symmetrised dot pattern technique generates images that one may use as the basis of a stall detection methodology [41]. One may process unsteady pressure or acoustic signals using the symmetrised dot pattern technique to generate distinctly different images at different fan speeds and operating conditions, Figure 12. The images which the technique produces are distinctly different during stable operation, when stall is incipient and during stalled operation.

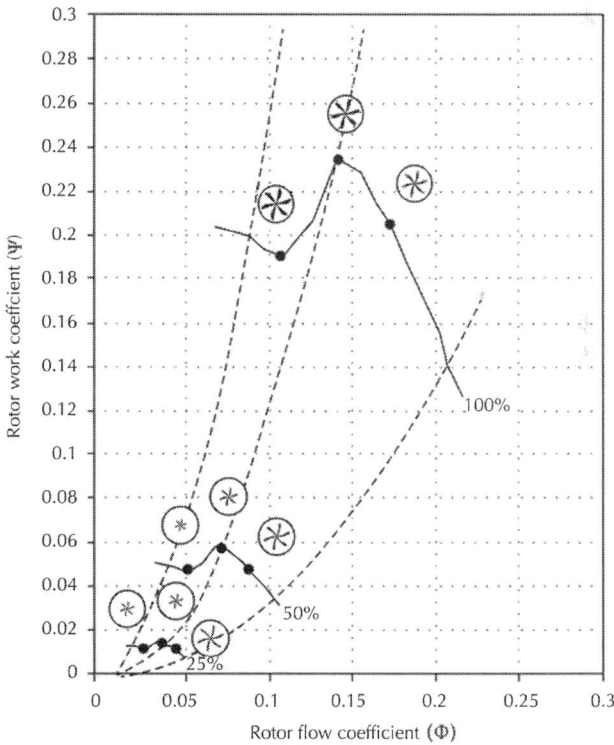

Figure 12: Unsteady pressure data from a 2.24 meter diameter tunnel ventilation fan at 100 per cent, 50 per cent, and 25 per cent speed in stable

operation, when stall is incipient and during stalled operation. The authors processed the data using the symmetrised dot pattern (SDP) technique to psroduce a set of patterns that are distinctly different at each speed and operating condition [41].

An ability to generate images sufficiently different to enable engineers to differentiate between stable operation, incipient stall and stalled operation results in a symmetrised dot pattern technique with the potential to form the basis of a stall detection system. Applying Bianchi et al.'s [71] symmetrised dot pattern technique to acoustic measurements constitutes a significant advance over the two-point spatial correlation, the stochastic model, travelling wave energy analysis or cross-correlation analysis as one can generate the images using a single acoustic signal. The other techniques require mounting a high frequency response pressure transducer in the industrial fan or compressor casing over the blades that are prone to stall. In contrast, one may apply the symmetrised dot pattern technique to acoustic measurement made with a single microphone at any location in close proximity to the industrial fan or compressor. The technique is particularly effective compared to other techniques when the signal of interest is low compared to the background noise [71]. This enables the symmetrised dot pattern technique to provide useful results when the microphone is situated in the acoustic far-field.

We can differentiate the symmetrised dot pattern technique from other stall detection systems as it can identify a shift from stable operation to incipient stall an order of magnitude more quickly than other techniques. Other stall detection techniques use a Fourier analysis to analyse raw pressure signals which generate the signal's frequency spectrum. They then identify a change in frequency spectrum as a fan moves from stable operation to incipient stall. A weakness associated with using a Fourier analysis is that the minimum sample size needed is relatively large compared to that required by the symmetrised dot pattern technique. Consequently, the other techniques require a longer data acquisition period that is required by the symmetrised dot pattern technique. The shorter the required data acquisition period needed for the signal analysis to be effective, the more likely that the resultant output will be available quickly enough to provide a warning that stall is incipient before a fan transitions into stalled operation.

We may illustrate the effectiveness of the symmetrised dot pattern technique when compared with any of the Fourier transform based stall detection techniques with an example. We present the output from a high frequency pressure transducer located in the far-field for a 20-second period during which the fan operating point moves from stable operation to incipient stall, Figure 13. In this context far-field refers to a pressure transducer located in the inlet box of an induced draft fan. The reason for using data from the far-field is that it extends stall detection from the hydrodynamic pressure near-field to the acoustic pressure far-field. In this example, we logged data at 2000 Hz. From zero to ten seconds the fan is operating in its operating characteristic's stable region. At ten seconds stall becomes incipient, and remains incipient for one second. From 11 to 20 seconds the fan is operating in its operating characteristic's unstable region.

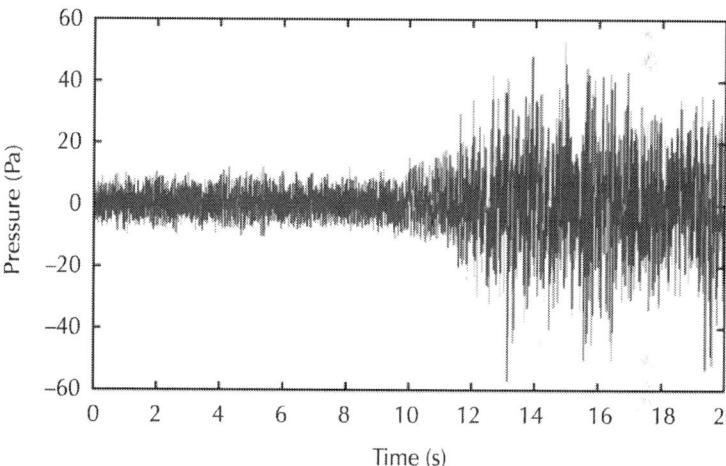

Figure 13: Data logged at 2000 Hz from a high frequency response pressure transducer located in the inlet box of an induced draft fan. From zero to ten seconds, the fan is operating in a stable condition. At ten seconds, stall becomes incipient. From ten to 20 seconds stall remains incipient.

A Fourier analysis of ten rotor revolutions of data (0.1 seconds) during stable operation, immediately before ten seconds, Figure 13, and ten rotor revolutions of data when stall is incipient, immediately after ten seconds, Figure 13, results in distinctly different frequency spectrum, Figure 14. This difference between the two frequency spectra below 50 Hz is a consequence of the spike-like pressure pulses associated with incipient stall present in the data from ten to 20 seconds, and absent in the data from zero to ten seconds. We used the difference between the frequency spectrum that we generated using the data from a stable operating condition and an operating condition where stall is incipient as the basis of Fourier transform based stall warning techniques.

Frequency spectrum for 10 rotor revolutions during stable operation

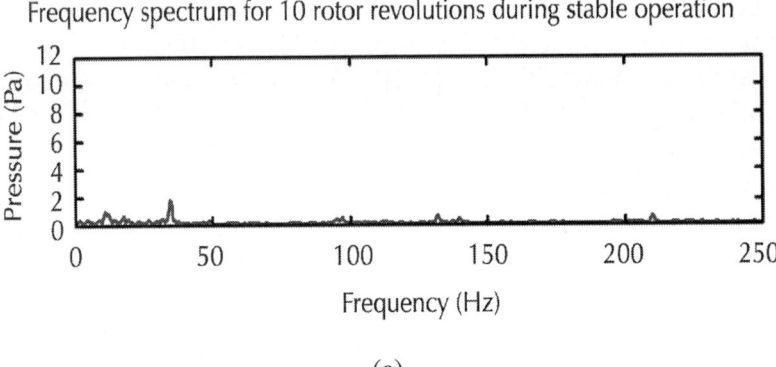

(a)

Frequency spectrum for 10 rotor revolutions during incipient stall

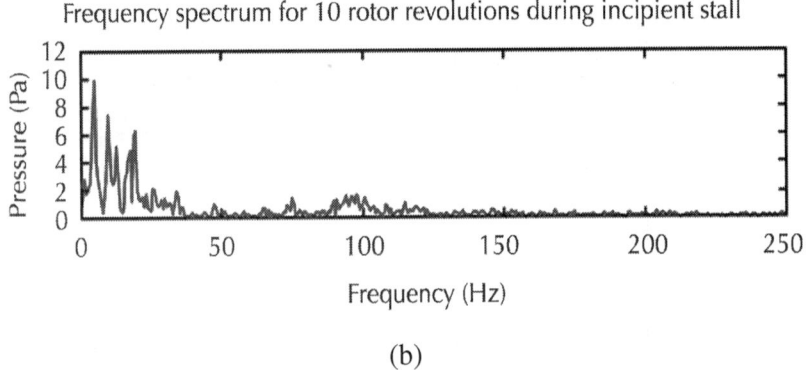

(b)

Figure 14: Frequency spectrum generated using data from over ten rotor revolutions during stable fan operation (top) and over ten rotor revolutions when stall is incipient (bottom). The frequency spectrum generated using data when stall is incipient (b) includes features that are associated with the spike-like pressure pulses that occur when stall is incipient. As such, the two frequency spectrums are different enabling one to use them as the basis of a stall detection system.

The signal processing associated with the symmetrised dot pattern technique does not involve a Fourier transform, instead of transforming the data into a set of polar coordinates that one uses to create the symmetrised dot patterns. In the above example, stall becomes incipient at ten seconds, Figure 13. One rotor revolution of data (0.01 seconds) during stable operation, immediately before ten seconds, Figure 13, and one rotor revolution of data when stall is incipient, immediately after ten seconds, Figure 13, results in distinctly different symmetrised dot patterns, Figure 15. It is the difference between the generated symmetrised dot pattern using the data from a stable operating condition and an operating condition where stall is incipient that enable one to use the symmetrised dot pattern technique as the basis of a stall warning technique.

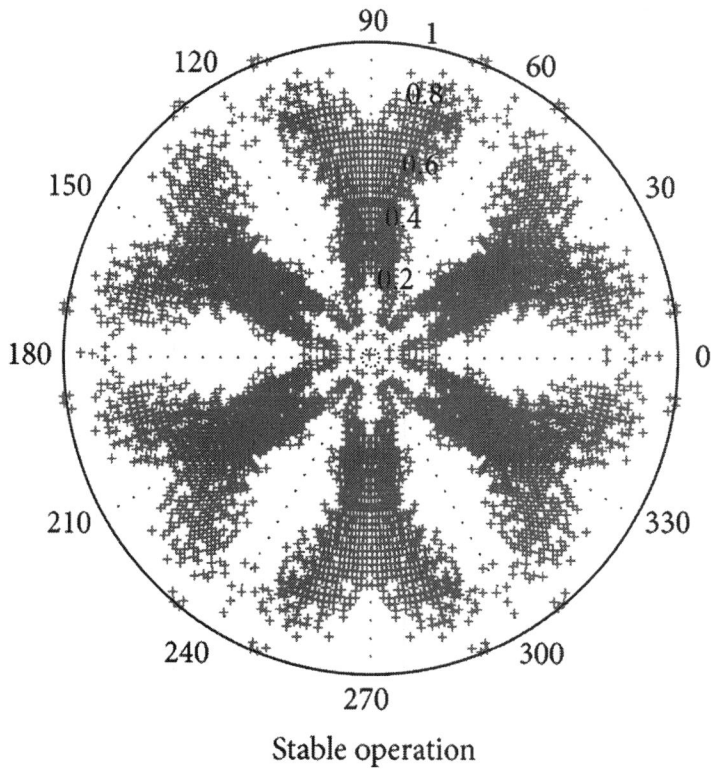

Stable operation

(a)

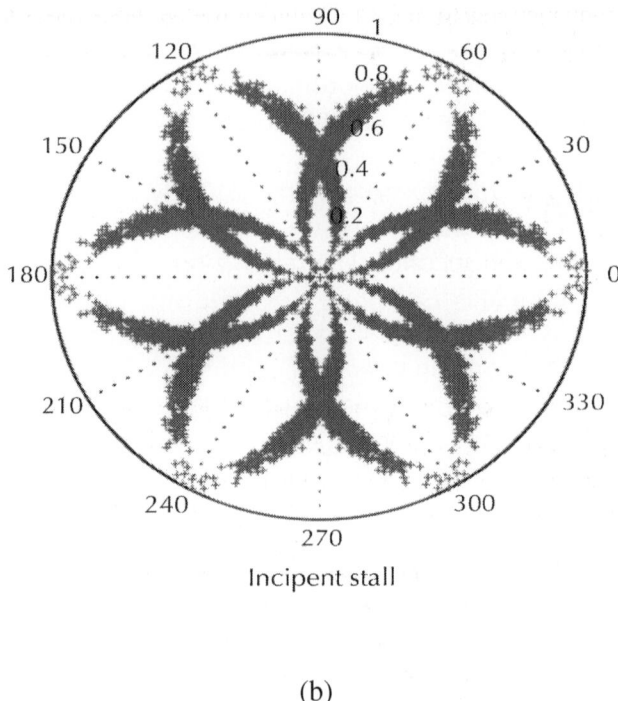

Incipent stall

(b)

Figure 15: Symmetrised dot patterns generated using data from over one rotor revolution during the fan's stable operation (a) and over one rotor revolution when stall is incipient (b). The two patterns are different enabling one to use them as the basis of a stall detection system.

When one conducts a Fourier analysis using the same one rotor revolution of data to generate each symmetrised dot pattern, the resultant frequency spectrum for the stable operating condition and an operating condition where stall is incipient are similar, Figure 16. A consequence of the frequency spectrum being similar is that any of the stall warning techniques based upon the use of a Fourier analysis will not be able to use the frequency spectrum to differentiate between a fan in stable operation and when stall is incipient. By contrast, the symmetrised dot pattern technique is able to generate distinctly different patterns, indicating that the symmetrised dot pattern technique can identify a change from a stable operating condition to one where stall is incipient an order of magnitude more rapid than stall warning techniques based on the use of a Fourier transform based analysis.

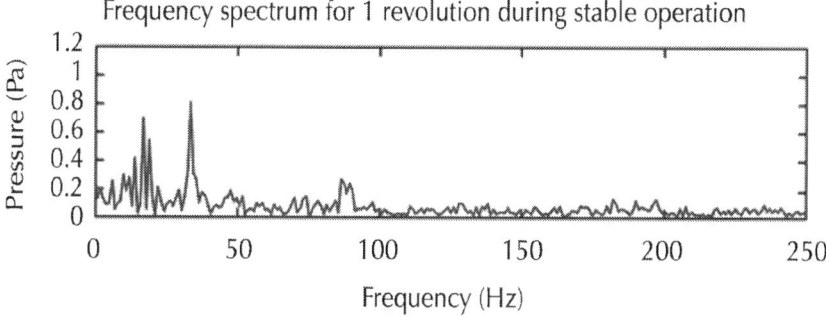

Figure 16: Frequency spectrum generated using data from over one rotor revolution during the fan's stable operation (top) and over one rotor revolution when stall is incipient (bottom). The two frequency spectrums are similar and therefore can't be used as the basis of a stall detection system.

The five stall detection techniques, Table 3, each represent a valid approach to stall detection. In practice it is the two-point spatial correlation technique that is the most developed stall detection technique in compressor applications and the symmetrised dot pattern technique that is the most widespread application in industrial fan applications. Stall detection systems well enough developed for in-service applications are still the subject of development in both the industrial fan and compressor communities. Research is ongoing in an effort to better understand the flow physics that occur with incipient stall in order to improve the accuracy and speed with which one can identify incipient stall.

Table 3: Matrix of stall detection techniques

Technique of detection	How it works	Results
Two-point spatial correlation	Finds spikes inception in the signal before the stall happens.	Valuable for analysis of prestall and stall inception [66].
Stochastic model	Correlates the pressure transducers' signals on the casing over the rotor tips. Finds dips in the correlation.	Used for stall detection [67].

Travelling-wave-energy analysis	Uses fan dynamics to introduce the concept of travelling wave energy as a real time measure of compressor stability.	Early warning for spikes; good for high speed compressors [68].
Cross-correlation analysis	Searches for short wavelength disturbances recognisable as spikes that form and decay before the stall.	Finds the spike inception, but does not use the instantaneous signal [69].
Acoustic stall detection	Expresses the time-series signal's changes in amplitude and frequency.	Discerns from critical and noncritical stall using an easy to understand visual waveform analysis [41].

CONCLUSIONS

This paper aimed to provide an overview of stall control technologies for industrial fans. These control technologies play an important role in many industrial applications. We have examined the stall phenomenon and paid particular attention to fluid dynamics, stall inception, and mechanical failure that may occur when operating industrial fans. We then introduced the technologies that exist today for stall identification and control, distinguishing between active and passive technologies. We can use passive technologies to prevent the worst stall consequence: mechanical failures. Industrial fan manufacturers primarily use passive control technologies in service.

Although innovation in passive technology is possible, passive technologies generally reduce fan efficiency. Current regulation within the European Union and planned regulation in the USA sets minimum Fan and Motor Efficiency Grades (FMEGs) for industrial fans. These minimum FMEGs will rise in the European Union on 1 January, 2015, and will continue to rise both within Europe and the USA in an on-going effort to reduce carbon emissions. As such, it is likely that the efficiency penalty with using passive technologies in industrial fan applications will become progressively less acceptable over the coming decade.

Researchers are focusing on active stall control technologies in an on-going effort to develop effective stall detection system. Active stall control technologies include technologies that are still the subject of both fundamental research and applied development and therefore, at the time of writing, are still immature. Despite the need for further research and development of active stall control systems, and their associated stall detection systems, they offer the greatest potential for medium term improvement. We can most readily realise the potential for medium term improvement in industrial fan applications, as industrial fans are relatively low speed machines in comparison to compressors. Consequently, active stall control systems based upon stall detection have the potential for practical application in industrial fans first, with the experience gaining in industrial fan application informing the development of higher speed systems that engineers can then use in compressor applications.

REFERENCES

1. B. de Jager, "Rotating stall and surge control: a survey," in Proceedings of the 34th IEEE Conference on Decision and Control, pp. 1857–1862, New Orleans, LA , USA, December 1995.

2. E. M. Greitzer, "Review—axial compressor stall phenomena," Journal of Fluids Engineering, Transactions of the ASME, vol. 102, no. 2, pp. 134–151, 1980.

3. I. J. Day and N. A. Cumpsty, "The measurement and interpretation of flow within rotating stall cells in axial compressors," Journal of Mechanical Engineering Science, vol. 20, no. 2, pp. 101–114, 1978.

4. F. K. Moore, "A theory of rotating stall of multistage compressors, parts I–III," Journal of Engineering for Gas Turbines and Power, vol. 106, no. 2, pp. 313–336, 1984.

5. H. W. Emmons, C. E. Pearson, and H. P. Grant, "Compressor surge and stall propagation," Transactions of the ASME, vol. 77, pp. 455–469, 1955.

6. J. T. Gravdahl and O. Egeland, Compressor Surge and Rotating Stall: Modeling and Control, Springer, London, UK, 1999.

7. A. G. Sheard and A. Corsini, "The mechanical impact of aerodynamic stall on tunnel ventilation fans,"International

Journal of Rotating Machinery, vol. 2012, Article ID 402763, 12 pages, 2012. · ·

8.   D. A. Fink, N. A. Cumpsty, and E. M. Greitzer, "Surge dynamics in a free-spool centrifugal compressor system," Journal of Turbomachinery, vol. 114, no. 2, pp. 321–332, 1992.

9.   A. M. Wo and J. P. Bons, "Flow physics leading to system instability in a centrifugal pump," Journal of Turbomachinery, vol. 116, no. 4, pp. 612–621, 1994.

10. J. D. Paduano, A. H. Epstein, L. Valavani, J. P. Longley, E. M. Greitzer, and G. R. Guenette, "Active control of rotating stall in a low-speed axial compressor," Journal of Turbomachinery, vol. 115, no. 1, pp. 48–57, 1993.

11. J. Parduano, L. Valavani, and A. H. Epstein, "Parameter identification of compressor dynamics during closed-loop operation," Journal of Dynamic Systems, Measurement and Control, vol. 115, no. 4, pp. 694–703, 1993.

12. J. D. Paduano, L. Valavani, A. H. Epstein, E. M. Greitzer, and G. R. Guenette, "Modeling for control of rotating stall," Automatica, vol. 30, no. 9, pp. 1357–1373, 1994. · ·

13. J. E. Pinsley, G. R. Guenette, A. H. Epstein, and E. M. Greitzer, "Active stabilization of centrifugal compressor surge," Journal of Turbomachinery, vol. 113, no. 4, pp. 723–732, 1991.

14. C. Rodgers, "Centrifugal compressor inlet guide vanes for increased surge margin," Journal of Turbomachinery, vol. 113, no. 4, pp. 696–702, 1991.

15. J. S. Simon and L. Valavani, "A Lyapunov based nonlinear control scheme for stabilizing a basic compression system using a close-coupled control valve," in Proceedings of the American Control Conference, vol. 3, pp. 2398–2406, June 1991.

16. J. S. Simon, L. Valavani, A. H. Epstein, and E. M. Greitzer, "Evaluation of approaches to active compressor surge stabilization," Journal of Turbomachinery, vol. 115, no. 1, pp. 57–67, 1993.

17. I. J. Day, "Axial compressor performance during surge," Journal of Propulsion and Power, vol. 10, no. 3, pp. 329–336, 1994.

18. G. Eisenlohr and H. Chladek, "Thermal tip clearance control for centrifugal compressor of an APU engine," Journal of Turbomachinery, vol. 116, no. 4, pp. 629–634, 1994.

19. A. H. Epstein, J. E. F. Williams, and E. M. Greitzer, "Active suppression of aerodynamic instabilities in turbomachines," Journal of Propulsion and Power, vol. 5, no. 2, pp. 204–211, 1989.

20. K. M. Eveker and C. N. Nett, "Model development for active surge control/rotating stall avoidance in aircraft gas turbine engines," in Proceedings of the American Control Conference, pp. 3166–3172, June 1991.

21. K. M. Eveker and C. N. Nett, "Control of compression system surge and rotating shell: a laboratory-based ‹hands-on› introduction," in Proceedings of the American Control Conference, vol. 2, pp. 1307–1311, June 1993.

22. K. M. Eveker, D. L. Gysling, C. N. Nett, and O. P. Sharma, "Integrated control of rotating stall and surge in aeroengines," in Sensing, Actuation, and Control in Aeropropulsion, pp. 21–35, April 1995.

23. J. E. F. Williams, M. F. L. Harper, and D. J. Allwright, "Active stabilization of compressor instability and surge in a working engine," Journal of Turbomachinery, vol. 115, no. 1, pp. 68–75, 1993.

24. J. E. F. Williams and X. Y. Huang, "Active stabilization of compressor surge," Journal of Fluid Mechanics, vol. 204, pp. 245–262, 1989.

25. A. Goto, "Suppression of mixed-flow pump instability and surge by the active alteration of impeller secondary flows," Journal of Turbomachinery, vol. 116, no. 4, pp. 621–628, 1994.

26. E. M. Greitzer and F. K. Moore, "A theory of post-stall transients in axial compression systems: part II—application," Journal of Engineering for Gas Turbines and Power, vol. 108, no. 2, pp. 231–239, 1986.

27. D. L. Gysling, M. Dugundji, J. E. Greitzer, and A. H. Epstein, "Dynamic control of centrifugal compressor surge using tailored structures," Journal of Turbomachinery, vol. 113, no. 4, pp. 710–722, 1991.

28. W. W. Copenhaver and T. H. Okiishi, "Rotating stall performance and recoverability of a high-speed 10-stage axial flow compressor," Journal of Propulsion and Power, vol. 9, no. 2, pp. 281–292, 1993.

29. I. J. Day, "Stall inception in axial flow compressors," Journal of Turbomachinery, vol. 115, no. 1, pp. 1–9, 1993.

30. K. H. Kim and S. Fleeter, "Compressor unsteady aerodynamic response to rotating stall and surge excitations," Journal of Propulsion and Power, vol. 10, no. 5, pp. 698–708, 1994.

31. S. Bianchi, A. Corsini, and A. G. Sheard, "Detection of stall regions in a low-speed axial fan, part 1: azimuthal acoustic measurements," in Proceedings of the 55th American Society of Mechanical Engineers Turbine and Aeroengine Congress, Glasgow, UK, Paper No. GT2010-22753, June 2010.

32. M. M. Bright, H. Qammar, H. Vhora, and M. Schaffer, "Rotating pip detection and stall warning in high-speed compressors using structure function," in Proceedings of the AGARD RTO AVT Conference, Toulouse, France, May 1998.

33. T. R. Camp and I. J. Day, "A study of spike and modal stall phenomena in a low-speed axial compressor," Journal of Turbomachinery, vol. 120, no. 3, pp. 393–401, 1998.

34. A. Deppe, H. Saathoff, and U. Stark, "Spike-type stall inception in axial flow compressors," inProceedings of the 6th Conference on Turbomachinery, Fluid Dynamics and Thermodynamics, pp. 178–188, Lille, France, 2005.

35. H. D. Vo, C. S. Tan, and E. M. Greitzer, "Criteria for spike initiated rotating stall," in Proceedings of the 50th American Society of Mechanical Engineers Gas Turbine and Aeroengine Congress, Reno, NV, USA, Paper No. GT2005-68374, June 2005.

36. S. Bianchi, A. Corsini, L. Mazzucco, L. Monteleone, and A. G. Sheard, "Stall inception, evolution and control in a low speed axial fan with variable pitch in motion," Journal of Engineering for Gas Turbines and Power, vol. 134, no. 4, Article ID 042602, 10 pages, 2012. ··

37. N. A. Cumpsty, "Part-circumference casing treatment and the effect on compressor stall," in Proceedings of the 34th American Society of Mechanical Engineers Gas Turbine and Aeroengine

Congress, Toronto, ON, Canada, Paper No. 89-GT-312, June 1989.

38. L. Mongeau, D. E. Thompson, and D. K. Mclaughlin, "A method for characterizing aerodynamic sound sources in turbomachines," Journal of Sound and Vibration, vol. 181, no. 3, pp. 369–389, 1995. · ·

39. K. Okada, "Experiences with flow-induced vibration and low frequency noise due to rotating stall of centrifugal fan," Journal of Low Frequency Noise and Vibration, vol. 6, no. 2, pp. 76–87, 1987.

40. F. Kameier and W. Neise, "Rotating blade flow instability as a source of noise in axial turbomachines,"Journal of Sound and Vibration, vol. 203, no. 5, pp. 833–853, 1997.

41. A. G. Sheard, A. Corsini, and S. Bianchi, "Detection of stall regions in a low-speed axial fan, part 2: stall warning by visualisation of sound signals," in Proceedings of the 55th American Society of Mechanical Engineers Turbine and Aeroengine Congress, pp. 14–18, Glasgow, UK, Paper No. GT2010-22754, June 2010.

42. A. Rippl, Experimentelle Untersuchungen Zuminstationaren Betriebsverhahenan der Stabilitarsgrenze Eines Mehrstufigen Transsonischen Verdichters [Ph.D. thesis], Ruhr-Universitat Bochum, 1995.

43. A. G. Sheard and N. M. Jones, "Powered smoke and heat exhaust ventilators: the impact of EN 12101-3 and ISO 21927-3," Tunnelling and Underground Space Technology, vol. 28, no. 1, pp. 174–182, 2012. · ·

44. EN12101-3, "Smoke and heat control systems. Specification for powered smoke and heat exhaust ventilators," 2002.

45. ISO and 21927-3, "Smoke and heat control systems—part 3: specification for powered smoke and heat exhaust ventilators.," 2006.

46. A. G. Sheard and A. Corsini, "The impact of an anti-stall stabilisation ring on industrial fan performance: implications for fan selection," in Proceedings of the 56th American Society of Mechanical Engineers Turbine and Aeroengine Congress, Vancouver, BC, Canada, Paper No. GT2011-45187, June 2011.

47.   D. Borello, A. Corsini, G. Delibra, F. Rispoli, and A. G. Sheard, "Numerical investigation on the aerodynamics of a tunnel ventilation fan during pressure pulses," in Proceedings of the 10th European Turbomachinery Conference, pp. 573–582, Lappeenranta, Finland, April, 2013.

48.   M. Gad-el-Hak, Flow Control: Passive, Active, and Reactive Flow Management, Cambridge University Press, Cambridge, UK, 2000.

49.   R. D. Joslin, R. H. Thomas, and M. M. Choudhari, "Synergism of flow and noise control technologies,"Progress in Aerospace Sciences, vol. 41, no. 5, pp. 363–417, 2005. · ·

50.   K. L. Suder, M. D. Hathaway, S. A. Thorp, A. J. Strazisar, and M. B. Bright, "Compressor stability enhancement using discrete tip injection," Journal of Turbomachinery, vol. 123, no. 1, pp. 14–23, 2001. · ·

51.   C. Nie, G. Xu, X. Cheng, and J. Chen, "Micro air injection and its unsteady response in a low-speed axial compressor," Journal of Turbomachinery, vol. 124, no. 4, pp. 572–579, 2002. · ·

52.   F. Lin, Z. Tong, S. Geng, J. Zhang, J. Chen, and C. Nie, "A summary of stall warning and suppression research with micro tip injection," in Proceedings of the 56th American Society of Mechanical Engineers Turbine and Aeroengine Congress, Vancouver, BC, Canada, Paper No. GT2011-46118, June 2011.

53.   H. J. Weigl, J. D. Paduano, L. G. Frechette et al., "Active stabilization of rotating stall and surge in a transonic single stage axial compressor," in Proceedings of the International Gas Turbine & Aeroengine Congress & Exposition, June 1997.

54.   H. D. Vo, J. Cameron, and S. Morris, "Control of short length-scale rotating stall inception on a high-speed axial compressor with plasma actuation," in Proceedings of the 53rd American Society of Mechanical Engineers Gas Turbine and Aeroengine Congress, Berlin, Germany, Paper No. GT2008-50967, June 2008.

55.   T. C. Corke and M. L. Post, "Overview of plasma flow control: concepts, optimization, and applications," in Proceedings of the 43rd AIAA Aerospace Sciences Meeting and Exhibit, Reno, NV, USA, Paper No. AIAA 2005-563, January 2005.

56. H. D. Vo, "Active suppression of rotating stall inception with distributed jet actuation," International Journal of Rotating Machinery, vol. 2007, Article ID 56808, 15 pages, 2007. · ·

57. J. V. R. Prasad, Y. Neumeier, M. Lal, S. H. Bae, and A. Meehan, "Experimental investigation of active and passive control of rotating stall in axial compressors," in Proceedings of the IEEE International Conference on Control Applications (CCA) and IEEE International Symposium on Computer Aided Control System Design (CACSD ‹99), pp. 985–990, August 1999.

58. S. Yeung and R. M. Murray, "Reduction of bleed valve rate requirements for control of rotating stall using continuous air injection," in Proceedings of the IEEE International Conference on Control Applications, pp. 683–690, October 1997.

59. C. S. Tan, I. Day, S. Morris, and A. Wadia, "Spike-type compressor stall inception, detection, and control," Annual Review of Fluid Mechanics, vol. 42, pp. 275–300, 2010. · ·

60. S. K. Ivanov, "Axial blower," US Patent, 3, 189–260, 1965.

61. S. Karlsson and T. Holmkvist, "Guide vane ring for a return flow passage in axial fans and a method of protecting it," US Patent 4, 602, 410, 1986.

62. T. Houghton and I. Day, "Enhancing the stability of subsonic compressors using casing grooves,"Journal of Turbomachinery, vol. 133, no. 2, Article ID 021007, 11 pages, 2011. · ·

63. N. Yamaguchi, M. Ogata, and Y. Kato, "Improvement of stalling characteristics of an axial-flow fan by radial-vaned air-separators nobuyuki yamaguchi," Journal of Turbomachinery, vol. 132, no. 2, Article ID 021015, 10 pages, 2010. · ·

64. A. R. Wadia, D. Christensen, and J. V. Prasad, "Compressor stability management in aircraft engines," in Proceedings of the 25th Congress of the International Council of the Aeronautical Sciences, ICAS, Hamburg, Germany, 2006-5.4.2, Paper No. 759, 2006.

65. D. Christensen, P. Cantin, D. Gutz et al., "Development and demonstration of a stability management system for gas turbine engines," Journal of Turbomachinery, vol. 130, no. 3, Article ID 031011, 9 pages, 2008. · ·

66. J. Cameron and S. Morris, "Spatial correlation based stall inception analysis," in Proceedings of the 52nd American Society of Mechanical Engineers Gas Turbine and Aeroengine Congress, pp. 14–17, Montreal, Canada, Paper No. GT2007-28268, May 2007.

67. M. Dhingra, Y. Neumeier, J. V. R. Prasad, A. Breeze-Stringfellow, H.-W. Shin, and P. N. Szucs, "A stochastic model for a compressor stability measure," Journal of Engineering for Gas Turbines and Power, vol. 129, no. 3, pp. 730–737, 2007. · ·

68. M. Tryfonidis, O. Etchevers, J. D. Paduano, A. H. Epstein, and G. J. Hendricks, "Prestall behavior of several high-speed compressors," Journal of Turbomachinery, vol. 117, no. 1, pp. 62–80, 1995.

69. H. G. Park, Unsteady disturbance structures in axial flow compressor stall inception [M.S. thesis], Massachusetts Institute of Technology, Cambridge, MA, USA, 1994.

70. S. Bianchi, A. Corsini, and A. G. Sheard, "Demonstration of a stall detection system for induced-draft fans," Journal of Power & Energy, 2013. ·

71. S. Bianchi, A. Corsini, and A. G. Sheard, "Experiments on the use of symmetrized dot patterns for in-service stall detection in industrial fans," Advances in Acoustic and Vibration, vol. 2013, Article ID 610407, 10 pages, 2013. ·

72. Eurovent1/11, Fans and System Stall: Problems and Solution, 2007.

73. H. Bard, "The stabilization of axial fan performance," in Proceedings of the Institution of Mechanical Engineers (IMechE) Conference C120/84 on the Installation Effects in Ducted Fan Systems, pp. 100–106, 1984.

Fault Diagnosis of Tennessee Eastman Process Using Signal Geometry Matching Technique

Han Li and De-yun Xiao

Department of Automation, Tsinghua University, 100084, Beijing, China

ABSTRACT

This article employs adaptive rank-order morphological filter to develop a pattern classification algorithm for fault diagnosis in benchmark chemical process: Tennessee Eastman process. Rank-order filtering possesses desirable properties of dealing with nonlinearities and preserving details in complex processes. Based on these benefits, the proposed algorithm achieves pattern matching through adopting one-dimensional adaptive rank-order morphological filter to process unrecognized signals under supervision of different standard signal patterns. The matching degree is characterized by the evaluation of error between standard signal and filter output signal. Initial parameter settings of the algorithm are subject to random choices and further tuned adaptively to make output approach standard signal as closely as possible. Data fusion technique is also utilized to combine diagnostic results from multiple sources. Different fault types in Tennessee Eastman process are studied to manifest the effectiveness and advantages of the proposed method. The results show that compared with many typical multivariate statistics based methods, the proposed algorithm performs better on the deterministic faults diagnosis.

INTRODUCTION

The last decades have been witnessing the modern large-scale processes developing toward high complexity and multiplicity in industries such as chemical, metallurgical, mechanical, logistics, andetc. These processes are generally characterized by a long-process flow with large operation scales and complicated mechanisms. The typical features are highly nonlinear, long-time delay, and heavily correlated among measurements [1]. Process monitoring, aiming to ensure that the operations satisfy the performance specifications and indicating anomalies, becomes a major challenge in practice. First, the requirements of process expertise for model-based methods often pose difficulties for operators not specializing in this realm; secondly, the system identification theory based methods need to postulate specified mathematical models, which are incapable of capturing varied nonlinearities. In addition, due to the growing number of sensors installed in processes, quantity of data constantly generated under different conditions soars by a few orders of magnitude or more compared to small-scale processes [2]. The fundamental dilemma for process monitoring is deficient knowledge to establish relative accurate mathematical process description while incomplete methodology to exploit abundant data to reveal process mechanisms and operational statuses. In large-scale processes, standard PI (proportional-integral) or PID (proportional-integral-derivative) closed-loop control schemes are often adopted to compensate for variable disturbances and outliers. However, excessive compensation may easily cause controllers overburden and a trivial glitch could eventually develop to catastrophic fault(s). Based on the considerations of practical limits, demands of safety operation, cost optimization as well as business opportunities in technical development, the problem of how to more effectively utilize mass amount of process data to meet the increasing demand of system reliability has received intensive attention of academics and practitioners in related areas. Among all the tasks, data-driven fault diagnosis, involving the use of data to detect and identify faults, is one of the most interesting research domains.

In previous extensively cited literature, Venkatasubramanian once proposed classical three subclasses of diagnostic techniques: quantitative model-based methods, qualitative model-based methods, and process history based method [3-5]. From a new perspective to further investigate Venkatasubramanian's classification, data-driven based fault diagnosis not only includes a large part of techniques in process history based method, but also some belonging to qualitative model-based methods. To view data-driven methods as an integrated type, we can re-divide fault diagnosis methods into three subclasses, namely analytical model-based methods, qualitative knowledge-based methods, and data-driven based methods (DDBM), where DDBM can be further divided into data transform based methods (DTBM) and data reasoning based methods (DRBM). Figure 1 illustrates the proposed classification. In general, DDBM are associated with the methods with insufficient information available to form mechanism model. These kinds of methods employ process data in dynamic system to perform fault detection, diagnosis, identification, and location. DTBM, to be more specifically, highlights the adoption of linear or nonlinear mathematical transforms to map original data to data in another form and the transforms are often reversible. The transformed data may be without clear physical meanings, but with more practicality. The key concept of data transform lies in two attributes: deterministic transform paradigm and realization of data compression. With this concept, the scope of DTBM is smaller and more concentrating compared to DDBM; the purpose for data utilization is more specific. DTBM also needs no in-depth knowledge about system structure as well as experience accumulation and reasoning knowledge which are necessary to DRBM. Besides, the implementation of DTBM algorithms are easily understood and realized, but the drawback may be less robust than model based methods. Dimension transformation (often dimension reduction), filtering, decomposition and nonlinear mapping are recognized as common tools for data transform.

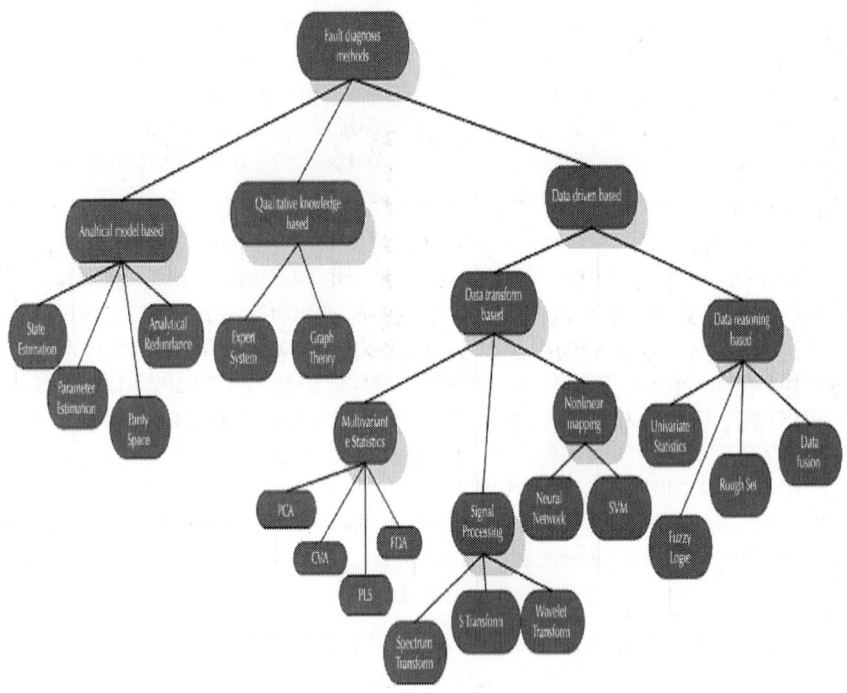

Figure 1: Classification of fault diagnosis methods proposed in this article.

In Figure 1, signal processing is categorized as a data transform methodology which covers a wide range of different techniques. Typical ones are primarily filtering and multilayer signal decomposition, both requiring preset models and carefully selected parameters, like Wavelet Analysis, Hilbert-Huang Transform, etc. Morphological signal processing, however, gives a different viewpoint. It derives from rank-order based data sorting technique and modifies signal geometry shape to achieve filtering [6]. This feature may provide more advantages of noise reduction and detail preservation than linear tools when treating measurements in complex processes [7]. Moreover, Salembier [8] analyzed that how the performance of rank-order based filter can be adaptively optimized in terms of the filter mask and rank value. Based on the investigations above, morphological signal processing as a nonlinear data transform tool may be suitable for constructing feature extractor for pattern matching.

In our previous work (unpublished work), we developed Salembier's idea [8] to adaptively adjust flat structuring element and rank parameter for each sample rather than adopting uniform ones for all the samples in a sampled sequence. Based on this idea, we designed a signal geometry matching approach: pattern classification using one-dimensional adaptive rank-order morphological filter for fault diagnosis, named PC1DARMF approach. The proposed method belongs to DTBM with major parameters capable of being randomly chosen, which is superior to those DTBM which need predefined parameters. This article applies PC1DARMF approach to a more complex and challenging application: Tennessee Eastman process (TEP). TEP is a classic model of an industrial chemical process widely studied in literature for validating new developed control or process monitoring strategies. It is a typical large-scale process characterized by features described previously. The fact that many data-driven diagnostic methods have been performed on TEP also provides chances to evaluate their performances in comparisons with method proposed in this article.

The remainder of this article is organized as follows: Section 2 expounds the derivation of pattern classification method using adaptive rank-order morphological filter. Key implementation issues are also discussed. An example is given to build a step-by-step realization of the method, making it easier for readers to understand. Section 3 gives an essential introduction to TEP and reviews the previous TEP fault diagnosis methods. Section 4 shows the diagnosis results for different TEP simulated faults with detailed analysis. Comparisons between the proposed method and typical multivariate statistics based approaches are made to highlight the advantages and features of PC1DARMF. The last part finally presents the conclusion and discussions.

SIGNAL GEOMETRY MATCHING BASED ON ADAPTIVE RANK-ORDER MORPHOLOGICAL FILTER

One-Dimensional Adaptive Rank-Order Morphological Filter (1DARMF)

Adaptive rank-order morphological filter is derived from a nonlinear signal processing tool referred as the rank-order based filter (ROBF). ROBF firstly reads a certain number of input values, then sorts the values in ascending order and determines the output value according to the predefined rank parameter in the sorted set. The basic definition of one-dimensional (1D) ROBF is firstly given in [9]: let x_i be discrete sampled signal defined on a 1D space Z and M be a 1D mask containing Npoints ($|M| = N$ and $|\ |$ is the set cardinality). Define j as an index belonging to the mask M and r as the normalized rank parameter of the filter ($0 \leq r \leq 1$). Given the rank-order operator denoted by $f_{r,M}[x_i]$, the output of ROBF y_i can be then formulated as (1):

$$y_i = f_{r,M}[x_i] = \text{Rank}_n\{x_{i-j} | j \in M\}$$

(1)

where elements of set X are sorted in ascending order and $\text{Rank}_n\{X\}$ denotes the nth ordered value in X (n is the nearest integer value of $(N - 1)r + 1$), xi-j denote all the points which belong to the range of mask M centered on i (e.g., if j = -3, -2, -1,0,1,2,3, i - j = i - 3,...,i+3). This operation is the essentials of both median filter and morphological filter with flat structuring element [8,9].However, its drawback is that the selections of filter mask and rank parameter heavily rely on practical experience and intuition. With understanding the feature of ROBF, its adaptive form named adaptive rank-order morphological filter was then proposed [8,9].It is optimized as adapting filter mask and rank parameter in order to minimize a criterion such as the MAE (mean absolute error) or the MSE (mean squared error). The problem of designing adaptive rank-order morphological filter can be briefly stated as follows: assume that x_i and d_i are given as noised signal and desired signal, respectively, when ROBF $f_{r,M}$ is adopted, the aim is to find the best rank parameter r and filter mask M which minimizes a cost function C between output y_i and d_i using iterative learning. In order to expound the procedure of building 1DARMF for better understanding, how to formulate the operation of ROBF is to be introduced at the beginning.

First, in order to overcome the optimization difficulty for dealing with the discrete nature of parameters, the rank parameter r can be optimized in continuous normalized manner and let n in $Rank_n\{X\}$ be the nearest integer value of $(N - 1)r + 1$. Secondly, for filter mask M optimization problem, a search area A which is selected to be larger than the optimum mask is introduced and a continuous value $m^{(j)}$ is assigned for $\forall j \in A$. New filter mask in next iterative step is thus determined by comparing the set of continuous values associated with the current filter mask against a preset value (denoted as threshold thm_M). If the assigned value for any $j \in A$ is greater than the threshold, the location associated to that j belongs to the filter mask. With introduction of search area A and the continuous values assignments, the optimization problem of filter mask M is successfully converted from the binary values modification of the mask (belong or not belong) to continuous values $m^{(j)}$ modification.

On the basis of realizing parameters updating continuously, we proceed to find a way to establish a mathematical relationship involving filter input, output, and the parameters all together. Let us define S the sum of signs of $(x_i\text{-}j\text{-}y_i)$ for all j. It can be expressed by

$$S = \sum_{j \in M} \text{sgn}\left(x_{i-j} - y_i\right)$$

(2)

It is easy to find out that if $r = 0$, y_i is the minimum of $\{x_i\text{-}j \mid j \in M\}$ and S is then equal to $N - 1$; if $r = 0.5$, y_i is the median value of $\{x_{i\text{-}j} \mid j \in M\}$ and $S = 0$; if $r = 1$, y_i is the maximum of $\{x_i j \mid j \in M\}$, $S = -(N - 1)$. Based on the mapping relations between S and r above, if they were assumed to be linearly related, the general expression of S with respect to r is given as

$$S = -(2r - 1)(N - 1)$$

(3)

In case of thm_M being set 0, we obtain if $(\text{sgn}(m^{(j)} - \text{thm\_M}) + 1)/2 = 1$, then $m^{(j)} > \text{thm\_M}$, which means $j \in M$ and else if $(\text{sgn}(m^{(j)} - \text{thm\_M}) + 1/2) = 0$, then $m^{(j) < \text{thm}}\text{-M}$, $j \in M^c$ Notice all j is selected from A and let $(\text{sgn}(m^{(j)} - \text{thm\_M}) + 1/2)$ (i.e., $(\text{sgn}(m^{(j)}) + 1)/2)$ be the weight, combing (2) and (3) gives

$$S = \sum_{j \in A} \frac{1}{2}(\text{sgn }(m^{(j)}) + 1)\text{sgn }(x_{i-j} - y_i) = -(2r - 1)[\sum_{j \in A}(\text{sgn }(m^{(j)}) + 1)/2 - 1]$$

(4)

$$F(m^{(j)}, x_{i-j}, y_j, r) = \sum_{j \in A} \frac{1}{2}(\text{sgn }(m^{(j)}) + 1)[\text{sgn }(x_{i-j} - y_i) + 2r - 1] + 1 - 2r = 0$$

(5)

Thus, the output of ROBF is successfully expressed by the implicit function $F(m^{(j)}, x_{i-j}, y_j, r)$. As will be stated later, this implicit function is applied to take derivatives of y_i with respect to m and r to develop iterative formulae for parameter updates.

In [8], an iterative algorithm similar to the LMS (least mean squares) algorithm was suggested to update the $m^{(j)}$ and r in the case of MSE optimization:

$$m^{(next,j)} = m^{(j)} + 2\alpha(d_i - y_i)\frac{\partial y_i}{\partial m^{(j)}} \quad \forall j \in A$$

(6)

$$r^{(next)} = r + 2\beta(d_i - y_i)\frac{\partial y_i}{\partial r}$$

(7)

Where   and   are two predefined parameters controlling the convergence rates. The derivatives of yj with respect to m(j) and r are calculated through employing implicit function (5). To obtain the expression of $\dfrac{\partial y_i}{\partial m_{(j)}}$ and $\dfrac{\partial y_i}{\partial r}$, the derivative of F with respect to mk is firstly expressed as

$$\frac{dF}{dm^{(j)}} = \frac{\partial F}{\partial m^{(j)}} + \left(\frac{\partial F}{\partial y_i}\right)\left(\frac{\partial y_i}{\partial m^{(j)}}\right) = 0$$

(8)

That is

$$\frac{\partial y_i}{\partial m^{(j)}} = -\frac{\partial F/\partial m^{(j)}}{\partial F/\partial y_i}$$

(9)

Using (5) to take the derivative of F with respect to $m^{(j)}$ gives

$$\frac{\partial F}{\partial m^{(j)}} = \frac{\partial \mathrm{sgn}\,(m^{(j)})}{2\partial m^{(j)}}[\mathrm{sgn}\,(x_{i-j} - y_i) + 2r - 1]$$
$$= \delta(m^{(j)})[\mathrm{sgn}\,(x_{i-j} - y_i) + 2r - 1]$$

(10)

$\dfrac{\partial F}{\partial y_i}$ is also calculated by using (5):

$$\frac{\partial F}{\partial y_i} = -\sum_{j\in A}(\mathrm{sgn}\,(m^{(j)}) + 1)\delta(x_{i-j} - y_j)$$

(11)

In (11), the term $\delta(x_{i-j}-y_i)$ is equal to 1 only if j equals to j_0, i.e., the time shift whose corresponding x_{i-j_0} equals to output y_i. This indicates $j_0 \in M$ and $\mathrm{sgn}(m_{j_0}) = 1$, (11) is simplified to

$$\frac{\partial F}{\partial y_i} = -2$$

(12)

Combined with (10), (9) is written as

$$\frac{\partial y_i}{\partial m^{(j)}} = \frac{1}{2}\delta(m^{(j)})[\mathrm{sgn}\ (x_{i-j} - y_i) + 2r - 1]$$

(13)

If $\delta(m_k)$ is replaced by $\delta_>(m_k) = 1$ for $-1 \leq m_k \leq 1$ for simplification. Based on (13), (6) is converted to

$$m^{(\mathrm{next},j)} = m^{(j)} + \alpha(d_i - y_i)[\mathrm{sgn}\ (x_{i-j} - y_i) + 2r - 1]$$

(14)

Similar with the deduction of (9) and (13), we also have

$$\frac{\partial y_i}{\partial r} = -\frac{\partial F/\partial r}{\partial F/\partial y_i}$$

(15)

$$\frac{\partial F}{\partial r} = 2\left[\frac{1}{2}\sum_{j \in A}(\mathrm{sgn}\ (m^{(j)}) + 1) - 1\right] = 2(N - 1)$$

(16)

Based on (12), (16) is written as

$$\frac{\partial y_i}{\partial r} = N - 1$$

(17)

Combined with (17), (7) is converted to

$$r^{(\text{next})} = r + 2\beta(d_i - y_i)(N - 1)$$

(18)

where $N = |M|$ is the current length of filter mask in use.

Combining (1), (14), and (18), the parameters updating algorithm for one dimensional adaptive rank order morphological filter are given as (19), where it N denotes the current iteration and it N + 1 for the next. Note that the update processes of filter mask M and rank parameter r are varying according to each sample i rather than remaining the same for each sample.

$$y_i^{(itN)} = \text{Rank}_{(N_i^{(itN)}-1)r_i^{(itN)}+1}\{x_{i-j}|j \in M_i^{(itN)}\}, |M_i^{(itN)}| = N_i^{(itN)}$$

$$\begin{cases} m_i^{(itN+1),j} = m_i^{(itN),j} + \alpha(d_i - y_i^{(itN)})[\text{sgn}\ (x_{i-j} - j) - y_i^{(itN)}) + 2r_i^{(itN)} - 1], \forall j \in M_i^{(itN)} \\ M_i^{(itN+1)} = \{j|\forall j \in M_i^{(itN)}, m_i^{(itN+1),j} > \text{thm\_M}\} \\ r_i^{(itN+1)} = r_i^{(itN)} + 2\beta(d_i - y_i^{(itN)})(N_i^{(itN)} - 1) \end{cases}$$

(19)

To illustrate the performance of 1DARMF given by (19), an example is shown in Figure 2. In Figure2a, it depicts three signals: noised signal x (dash-dot line) as input signal, desired signal d (solidline) as supervisory signal, and output signal y (dotted line) as recovered signal. $x = s + n$, wheres is the useful signal contaminated by Gaussian noise n and SNR_x (signal-to-noise ratio) is set 2. In this example, $s = \sin(t)$ and d is selected equal to s in order to recover the useful signal. Initial parameters of 1DARMF in (19) are set as follows: initial 1D filter mask $M^{(0)} = [-5,-4,-3,-2,-1,0,1,2,3,4,5]$, initial assigned value for element in the mask $m^{(0,j)} = 0.5$ ($\forall j \in M$), initial rank parameter $r^{(0)} = 0$, thm_M = 0, max iterations iterationNUM = 300, convergence rate $\alpha = 1 \times 10^{-4}$ and $\beta = 1.5 \times 10^{-3}$.

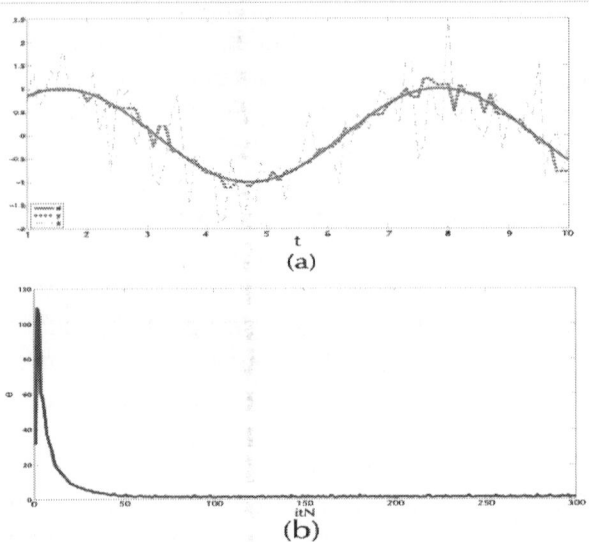

Figure 2: An example illustrating the performance of 1DARMF given by (19): (a) Supervisory signal d, noised signal x and output signaly and (b) $e^{(itN)}$ defined in (20).

If we define the sum of squared error between y and d as the evaluation of signal recovering ability, the expression is given as

$$e^{(itN)} = \sum_i |y_i^{(itN)} - d_i|^2$$

(20)

where i means the ith sample of signal and it N denotes current iteration. Figure 2b shows $e^{(itN)}$ converges to steady state and oscillates in a stable manner as it N gets increased.

Pattern Classification Using 1DARMF (PC1DARMF)

In Section 2.1, the general procedure to implement 1DARMF needs desired signal d as supervisory signal to train the key parameters of filter to obtain desired output. However, for a certain input x, if d is alternatively chosen, the iterative training process would finally lead to different output y. This means under supervision of inappropriate or undesirable d, the output may fail to recover useful signal from original input x. A performance comparison of 1DARMF using different supervisory signals is given to illustrate this phenomenon in Figure 3. With input x and the initial parameters being set the same with Section 2.1, different d results in different y, as shown in Figure 3a, c, e, g, i. Figure 3b, d, f, h, j depict corresponding $e^{(itN)}$ gradually reaches stable oscillation as iterations increase. The most distinct common feature is all $e^{(itN)}$ eventually progress to a steady-state through enough iterations. This phenomenon can be theoretically guaranteed: Feuer and Weinstein [10] concluded that if the convergence rate was restrained within a upper limit, then it was the necessary and sufficient for LMS algorithm to ensure the convergence of the algorithm. Therefore, with the proper selection of α in (6) and β in (7), $e^{(itN)}$ is also expected to stably oscillate eventually. The selection rule will be later summarized in Section 2.3. This condition is the crucial prerequisite to further form our algorithm for pattern classification. In Table 1 min($e^{(itN)}$) are also listed to numerically compare the effect of different d on signal recovering.

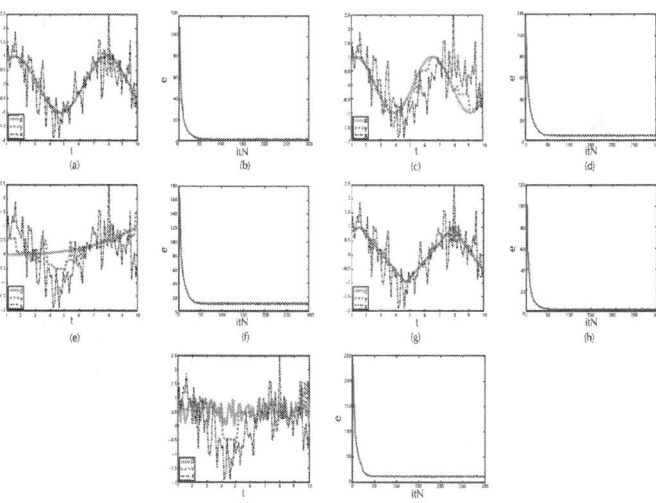

Figure 3: 1DARMF performances using s = sin(t), SNR_x = 2 and different supervisory signal d. Initial parameter settings: $M^{(0)}$ = [-5,-4,-3,-2,-1,0,1,2,3,4,5],

$m^{(0,j)} = 0.5$ ($\forall j \in B$), $r^{(0)} = 0$, thm_M = 0, iterationNUM = 300; (a) d = sin(t), (c) d = sin(1.2t), (e) d = c(t^3 + t^2 - 1)(c is a proper scaling factor which constrains range of d to be within [-1,1]), (g) d is triangular signal (TriWave), (i) d is signal generated according to uniform distribution (rand),(b), (d), (f), (h), (j): correspondent $e^{(itN)}$ of its left figure.

Table 1: min($e^{(itN)}$) gained using different supervisory signal d (s = sin(t))

S	D	min(e(itN))
sin(t)	sin(t)	0.7276
	sin(1.2t)	3.7734
	c(t3+ t2- 1)	8.9434
	TriWave	0.9754
	Rand	10.6224

Li and Xiao EURASIP Journal on Advances in Signal Processing 2011 2011:83 doi:10.1186/1687-6180-2011-83

Figure 3 and Table 1 indicate the most matching supervisory signal in signal geometry shape with original input x (i.e., d = s = sin(t)) yields minimum value of min($e^{(itN)}$), showing the best signal recovering ability. Based on this property, it is expected that given an unrecognized noised signal and a certain number of reference signals (also known as signal templates) as supervisory signals, 1DARMF may be capable of achieving signal recognition and classification through finding out under which reference signal the min($e^{(itN)}$) value reach the minimum among all reference signals provided. We thus propose the basic procedures for pattern classification using 1DARMF in Figure4.

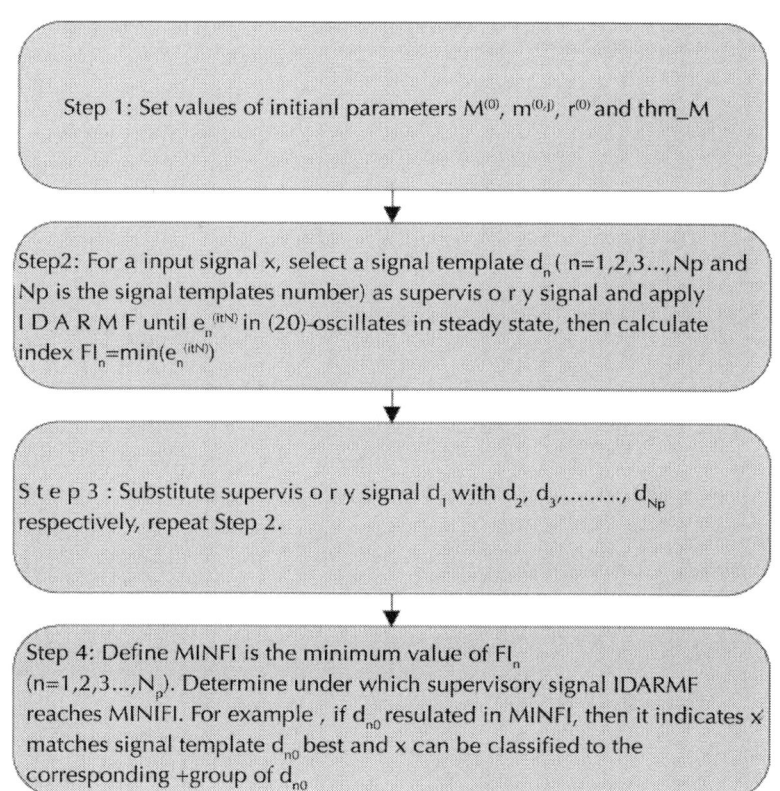

Figure 4: The framework of pattern classification using 1DARMF.

The procedure for pattern classification using 1DARMF can be further developed to an algorithm, named PC1DARMF algorithm. It is a supervised pattern classification approach. The fundamental of this algorithm is to realize signal geometry shape matching using 1DARMF as a tool in an iterative way. If the supervisory signals denote different types of physical meanings, for example representing different operation conditions or fault types in dynamic processes, this algorithm could achieve faults diagnosis through the signal geometry shape matching. In general, PC1DARMF algorithm is meaningful in two levels: first, it serves for the type classification purpose and secondly a feature extractor from nonstationary signals with proper parameter settings.

Issues for Implementing PC1DARMF Algorithm

In Section 2.2, PC1DARMF algorithm was mainly described in a high-level structure. There are still several significant engineering principles and experience to know which are important to practical implementation. They include initial parameter settings, convergence rates selections, and iteration stopping criteria.

Initial Parameter Settings

Initial parameter settings for PC1DARMF algorithm involves initial value determination of filter mask $M^{(0)}$, assigned value $m^{(0,j)}$ for each element in filter mask, rank parameter $r^{(0)}$ and the threshold thm_M. Several reasons are supporting the random initial parameter settings. First, the only variable of filter mask in 1DARMF is its length. Based on analysis of Nikolaou and Antoniadis[11] of empirical rule for the length selection and consideration of keeping computational complexity relatively low, we propose to random chose it between 0.3 and 0.5 times of the total length of input signal. Secondly, there are no guidelines in theory for m_i and r_i initial values. They get renewal in continuous manner to optimal value during iterations, so their initial values are expected to be different chosen each time within an interval (e.g., [0, 1]). Thirdly, notice the derivations of (6) and (18) in Section 2.1 are all irrelevant to the value of thm_M, thm_M can be also randomly chosen within [0, 1]. Besides, the most important is that it is impossible to find optimal initial parameter settings for signals with varying nonstationary characteristics. The first goal of PC1DARMF is to measure how good two signals match each other rather than achieve optimal signal recovering, so the selection of initial parameter values would not be necessarily restrained as special ones. Based on the analysis, we use random initial parameter settings for later experiments.

Convergence Rates Selections

The selection rule of convergence rate α and β in (19) is (21), which is referenced from [10] and early mentioned in Section 2.1. As was indicated before, (21) guarantees the convergence of the LMS algorithm.

$$0 < \mu \leq \frac{1}{3tr[R]}$$

(21)

where μ denotes convergence rate, R is covariance matrix of input signal, tr[R] is the trace of R. We further find empirically that if α and β is chosen as 1/3tr[R], output y may often cause unstable oscillation. In this article, we adopt that α and β is much smaller than 1/3tr[R]: for example, $\alpha = 0.0001, \beta = 0.0015$.

Iteration Stop Criteria

Max iteration number preset is the key factor to greatly influence the algorithm computational cost. Notice the computational complexity of PC1DARMF algorithm is $O(|\bar{N}\log\bar{N}||SL||dNUM||MaxitN|)$, where $\bar{N}$ is the average length of structuring element and $O(|\bar{N}\log\bar{N}|)$ is the computational complexity of Quicksort algorithm, SL is the processed signal length, and dNUM for the number of signal templates. SL and dNUM are predefined and unchangeable. MaxitN is the max iterations to ensure the convergence. Salembier [8] and Figure 3 in Section 2.2 also pointed out that 1DARMF had an ability to provide fast convergence. If the PC1DARMF algorithm always set a fixed iteration numbers, it would be unnecessary and the computational cost would be tremendous. An alternative way for reducing redundant iterations is to stop the iterations when within a certain number of continuous iterations, average variation of e^{itN} falls below a threshold if no specified information about input signal and the noise level is given.

TENNESSEE EASTMAN PROCESS FAULT DIAGNOSIS USING PC1DARMF ALGORITHM

Introduction to Tennessee Eastman Process (TEP)

Tennessee Eastman process is first proposed by Downs and Vogel [12] to provide a simulated model of real industrial complex process for studying large-scale process control and monitoring methods. As is shown in Figure 5, the process consists of five major units: an exothermic two-phase reactor, a product condenser, a recycle compressor, a flash separator, and a reboiler stripper. Gaseous reactants A, C, D, E, and inert B are fed to the reactor. Component G and H are two products of TEP, while F is undesired byproduct. The reaction stoichiometry is listed as (22). All the reactions are irreversible, exothermic, and approximately first-order with respect to the reactant concentrations. The reaction rates are expressed as Arrhenius function of temperature. The reaction producing G has higher activation energy than that producing H, thus resulting in more sensitivity to temperature.

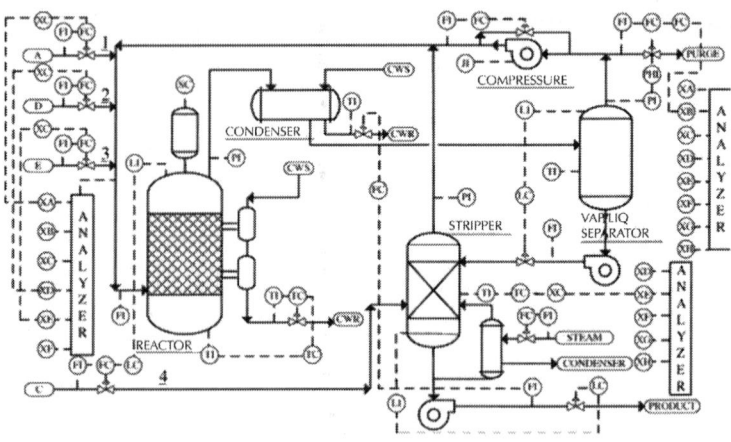

Figure 5: TEP flowsheet adopting control structure proposed by[13] .

$$A_{(g)} + C_{(g)} + D_{(g)} \rightarrow G_{(l)}$$
$$A_{(g)} + C_{(g)} + E_{(g)} \rightarrow H_{(l)}$$
$$A_{(g)} + E_{(g)} \rightarrow F_{(l)}$$
$$3D_{(g)} \rightarrow 2F_{(l)}$$

$$(22)$$

The reactor product stream is cooled through a condenser and fed to a vapor-liquid separator. The vapor exits the separator and recycles to the reactor feed through a compressor. A portion of the recycle stream is purged to prevent the inert and byproduct from accumulating. The condensed component from the separator is sent to a stripper, which is used to strip the remaining reactants. After G and H exit the base of the stripper, they are sent to a downstream process which is not included in the diagram. The inert and byproducts are finally purged as vapor from vapor-liquid separator.

The process provides 41 measured and 12 manipulated variables, denoted as XMEAS(1) to XMEAS(41) and XMV(1) to XMV(12), respectively. Their brief descriptions and units are listed in Table 2. Twenty preprogrammed faults IDV(1) to IDV(20) plus normal operation IDV(0) of TEP are given to represent different conditions of the process operation, as listed in Table 3. TEP proposed in [12] is open loop unstable and it should be operated under closed loop. Lyman and Georgakis[13] proposed a plant-wide control scheme for the process. In this article, we implement this control structure to evaluate performance of PC1DARMF algorithm on fault diagnosis for it provides the best performance for the process.

Table 2: Measurements and manipulated variables in TEP

Variable	Description	Units
XMEAS(1)	A feed (Stream 1)	kscmh
XMEAS(2)	D feed (Stream 2)	kg/h
XMEAS(3)	E feed (Stream 3)	kg/h
XMEAS(4)	Total feed (Stream 4)	kscm h

XMEAS(5)	Recycle flow (Stream 8)	kscm h
XMEAS(6)	Reactor feed rate (Stream 6)	kscm h
XMEAS(7)	Reactor pressure	kPa gauge
XMEAS(8)	Reactor level	%
XMEAS(9)	Reactor temperature	°C
XMEAS(10)	Purge rate (Stream 9)	kscm h
XMEAS(11)	Product sep temp	°C
XMEAS(12)	Product sep level	%
XMEAS(13)	Prod sep pressure	kPa gauge
XMEAS(14)	Prod sep underflow (Stream 10)	m3/h
XMEAS(15)	Stripper level	%
XMEAS(16)	Stripper pressure	kPa gauge
XMEAS(17)	Stripper underflow (Stream 11)	m3/h
XMEAS(18)	Stripper temperature	°C
XMEAS(19)	Stripper steam flow	kg/h
XMEAS(20)	Compressor work	kW
XMEAS(21)	Reactor cooling water outlet temp	°C
XMEAS(22)	Separator cooling water outlet temp	°C
Variable	Description	Stream
XMEAS(23)	Component A	6
XMEAS(24)	Component B	6
XMEAS(25)	Component C	6
XMEAS(26)	Component D	6
XMEAS(27)	Component E	6
XMEAS(28)	Component F	6
XMEAS(29)	Component A	9
XMEAS(30)	Component B	9
XMEAS(31)	Component C	9
XMEAS(32)	Component D	9
XMEAS(33)	Component E	9
XMEAS(34)	Component F	9
XMEAS(35)	Component G	9
XMEAS(36)	Component H	9
XMEAS(37)	Component D	11
XMEAS(38)	Component E	11

XMEAS(39)	Component F	11
XMEAS(40)	Component G	11
XMEAS(41)	Component H	11
Variable	Description	
XMV(1)	D feed flow (Stream 2)	
XMV(2)	E feed flow (Stream 3)	
XMV(3)	A feed flow (Stream 1)	
XMV(4)	Total feed flow (Stream 4)	
XMV(5)	Compressor recycle valve	
XMV(6)	Purge valve (Stream 9)	
XMV(7)	Separator pot liquid flow (Stream 10)	
XMV(8)	Stripper liquid product flow (Stream 11)	
XMV(9)	Stripper steam valve	
XMV(10)	Reactor cooling water flow	
XMV(11)	Condenser cooling water flow	
XMV(12)	Agitator speed	

Li and Xiao EURASIP Journal on Advances in Signal Processing 2011 2011:83 doi:10.1186/1687-6180-2011-83

Table 3: Notations and descriptions of faults in TEP

Variable	Description	Type
IDV(0)	Normal operation	-
IDV(1)	A/C feed ratio, B composition constant (Stream 4)	Step
IDV(2)	B composition, A/C ratio constant (Stream 4)	Step
IDV(3)	D feed temperature (Stream 2)	Step
IDV(4)	Reactor cooling water inlet temperature	Step
IDV(5)	Condenser cooling water inlet temperature	Step
IDV(6)	A feed loss (Stream 1)	Step
IDV(7)	C header pressure loss-reduced availablity (Stream 4)	Step

IDV(8)	A, B, C feed composition (Stream 4)	Random Variation
IDV(9)	D feed temperature (Stream 2)	Random Variation
IDV(10)	C feed temperature (Stream 4)	Random Variation
IDV(11)	Reactor cooling water inlet temperature	Random Variation
IDV(12)	Condenser cooling water inlet temperature	Random Variation
IDV(13)	Reaction kinetics	Slow Drift
IDV(14)	Reactor cooling water valve	Sticking
IDV(15)	Condenser cooling water valve	Sticking
IDV(16)	Unknown	
IDV(17)	Unknown	
IDV(18)	Unknown	
IDV(19)	Unknown	
IDV(20)	Unknown	

Li and Xiao EURASIP Journal on Advances in Signal Processing 2011 2011:83 doi:10.1186/1687-6180-2011-83

Related Work for TEP Fault Diagnosis

Various approaches have been proposed to deal with the fault diagnosis and isolation for TEP since its introduction in 1993. Most of them are dedicated to exploit data-driven techniques because of the process complexity and data abundance. Multivariate statistics based, machine learning based, and pattern matching based methods are the most frequently adopted methodologies summarized in this article. Meanwhile hybrids of the three have been also studied in literature.

Raich and Cinar [14-16] are among the earliest researchers to apply multivariate statistics techniques for TEP fault diagnosis. Training data under different operation conditions are firstly utilized to design PCA (principal component analysis) models for fault detection and fault classification. Then, designed PCA models are applied to new data to calculate statistic metrics and different discriminant analysis is conducted to determine whether and which fault occurs. The method is also able to diagnosis multiple simultaneous disturbances by quantitatively measuring the similarities between models for different fault types. Russell et al. [17] and Chiang et al. [18] gives a comprehensive and detailed study of multivariate statistical process monitoring using major dimensionality reduction techniques: PCA, FDA (Fisher discriminant analysis), PLS (partial least squares), and CVA (canonical variate analysis). Additionally, some improved multivariate statistical methods outperform their conventional counterparts for TEP fault diagnosis, like dynamic PCA/FDA (DPCA/DFDA) [19], moving PCA (MPCA) [20], and modified independent component analysis (modified ICA) [21]. Application of the multivariate statistics based methods is under assumption that sample data mean and covariance are equal to their actual values [17]. This would leads to requirement of large quantity of real data for ensuring relative accurate statistic estimations.

Machine learning based methods are also abundant in literature. It requires large amount of historical data under various fault conditions as training data to form a data mapping mechanism. Artificial neural networks (ANN) and support vector machine (SVM) are the most employed techniques applied to TEP fault diagnosis [22-25] among machine learning based methods. Eslamloueyan [26] further proposed hierarchical artificial neural network (HANN) to diagnosis faults for TEP. Fault pattern space is first divided to subspaces using fuzzy clustering algorithm. For each subspace representing a fault pattern, a special NN is trained for fault diagnosis. Besides, Bayesian networks [27,28] and signed directed graphs (SDG) [29] are also investigated in TEP fault diagnosis problem.

Another important approach is pattern matching. The basic idea is to match the pattern against the templates stored after using feature extracting techniques. Different similarity measures are defined to quantify the matching degree. Qualitative trend analysis (QTA) is a significant pattern-matching based method. It represents signals as a set of basic shapes as major features, which distinguishes different signals in geometry shapes. Maurya et al. [30] used seven primitives to represent signal geometry under different fault conditions. Maurya et al. [31] also proposed an interval-halving method for trend extraction and a fuzzy matching based method for similarity estimation and inferences. Akbarya and bishnoi [32] used wavelet-based method to extract features and binary decision tree to classify them. All the above, QTA-based methods require training data, while Singhal and Seborg [33] proposed a pattern-matching-strategy requires no training data but a huge amount of historical data. The approach needs specification of snapshot dataset, which serves as a template during the historical database search. Pattern similar to snapshot data in historical database can be located by sliding a window of signals in fixed length. The drawback of this method is that it needs to accumulate historical data and, of course, cannot perform on-line process monitoring tasks. In general, pattern recognition based methods are relatively computationally demanding but the development of computer processor has been helping lessen this pressure.

Hybrid TEP fault diagnosis methods, investigated in literature commonly, employ multivariate statistical tools. For example, Lee et al. [34] combined SDG and PLS to demonstrate better diagnosis resolution, accuracy, and reliability than previous qualitative methods. Lu et al. [35] considered the limitation of PCA for differentiating faults with similar time-varying characteristics and utilized wavelet analysis to extend the feature extracting ability into time-frequency domain.

PC1DARMF algorithm in this article is a novel pattern matching method. The theoretical basis is rank-order based filter theory, which is easy to understand and implement. The computational complexity is controllable and may be relatively lower than traditional QTA-based methods. It employs the complete signal rather than some elements extracted as template. The applied strategy preserves important information, preventing possible information loss, and distortion. It also needs no knowledge about occurrence moments of the fault. Simulation data in Section 4 would verify its effectiveness.

Diagnostic Procedure of Using PC1DARMF Algorithm

The method proposed for TEP fault diagnosis is a supervised signal geometry shape matching approach, so constructing the signal templates as supervisory signals should be considered in the first place. The left part of Figure 6 depicts the procedure for obtaining the templates. The training data used for template construction is a matrix consisting of raw sampled signal intervals of selected measurements within different fault conditions. The normalization of the raw signal to zero mean, unit variance signal eliminates the discrepancy in the different weights given to different variables. After noise reduction by wavelet de-noising method [36], PCA model is then employed to extract PCs (principal components) since variables in TEP are highly correlated. Signals of selected PCs are defined as signal templates. This means each fault (including normal operation) is now represented as a number of signal templates. When new sampled signals as testing data are available, PCA model for training data is applied to normalized and de-noised testing signals. PC1DARMF algorithm matches unrecognized signal patterns against every template to give classification result according to each PC. Finally, multi-data fusion technique, for instance consensus theory [37], combines the classification results of selected PCs to give the final decision.

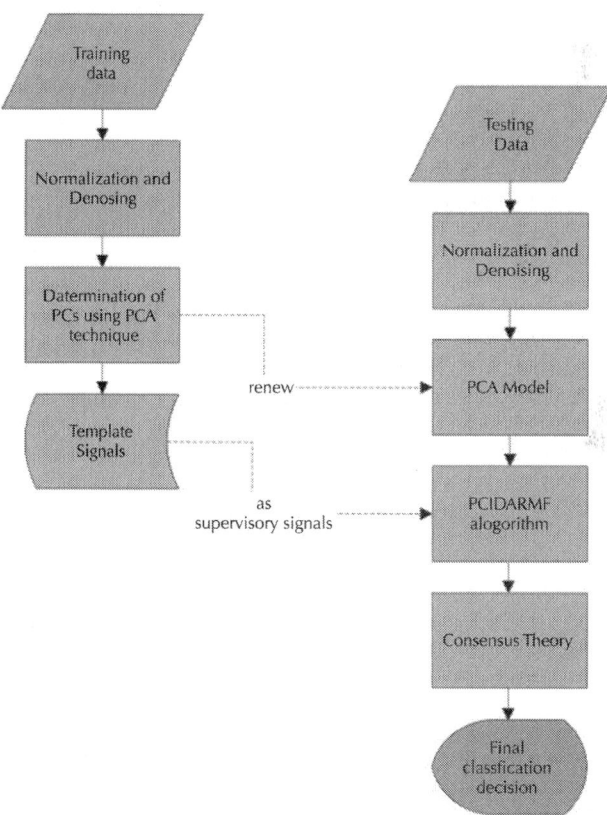

Figure 6: TEP fault diagnosis procedure using PC1DARMF algorithm.

SIMULATION RESULT ANALYSIS

Data Set Specification for Simulation

This section describes TEP simulation data specification we adopt in this article. It mainly concerns the data constituent for the training and testing sets, data sampling interval, and sample size. The form of training and testing data is a matrix consisting of variables XMEAS(1) to XMEAS(2) and XMV(1) to XMV(11) except constant-valued XMV(12). An observation vector of TEP process is given as (23) and data collected during one simulation run of fault type i consists of k observations assembled as (24). Downs and Vogel [12] recommended the favorable time for one simulation run was between 24 and 48 hrs. In this article, we chose 24 h to keep computational cost relatively low. Russell et al. [17] proposed a sampling interval of 3 min to allow fast fault diagnosis. Thus, one simulation run contains 480 observations in our simulation studies, i.e., k are 480 in (24). If n_i simulation runs are implemented for fault type i, the training data matrix including n_c fault types is represented as (25).

$$x = [XMEAS(1), ..., XMEAS(41), XMV(1), ..., XMV(11)]^T \tag{23}$$

$$X_i = \begin{bmatrix} x_{i1}^T \\ ... \\ x_{ik}^T \end{bmatrix} \tag{24}$$

$$M_{tr} = [X_{11}, ..., X_{1n_1}, ..., X_{i1}, ..., X_{in_i}, ..., X_{n_c 1}, ..., X_{n_c n_{n_c}}]^T \tag{25}$$

Deterministic Fault Diagnosis in TEP

In this section, we first consider diagnosis of deterministic faults, i.e., IDV(1) to IDV(7) in TEP. Their brief introductions are listed in Table 3. Following the procedures described in Section 3.3, we construct the signal templates as standard patterns in the first place. The training data matrix M_{tr} of (25) is assembled. It is designed to contain process data produced in eight different simulation runs and each fault type corresponds to a simulation. In these simulations, the fault is introduced 8 h after the simulation started. After PCA is conducted on normalized and de-noised M_{tr}, parallel analysis [38] is applied to suggest that the PCs, which correspond to the first five largest eigenvalues of training data covariance matrix, capture total variations of the data set optimally. To form standard patterns of each fault type, matrices X_0 to X_7 defined in (24) are first normalized and de-noised, respectively. After that projections of preprocessed data onto the first five PCA loading vectors are performed to obtain five new observations (scores). With the consideration of reducing the computational cost of later PC1DARMF algorithm, a resampling of the new observations reducing data points from 480 to 60 is performed. Figure 7 illustrates trends of five resampled signals for each fault type. With five in a row composing a set, these trends obtained from training data set are signal templates.

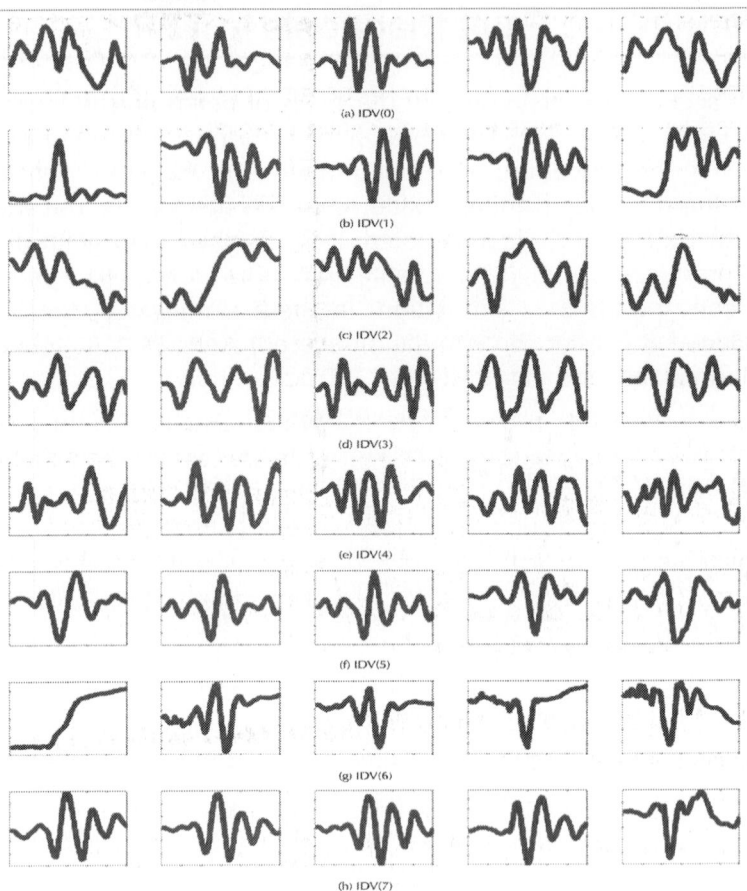

(a) IDV(0)

(b) IDV(1)

(c) IDV(2)

(d) IDV(3)

(e) IDV(4)

(f) IDV(5)

(g) IDV(6)

(h) IDV(7)

Figure 7: Set of signal templates for (a) IDV(0) to (h) IDV(7). In each set, five figures from left to right correspond to PC_1 to PC_5, respectively.

After signal templates were prepared, we move to build signal patterns derived from testing data. In Section 4.2, we define testing data set for each fault type, which consists of data collected from 20 simulation runs, in which 10 are generated with fault introduced in the 4th h and the 10 others introduced in the 12th h. As the same with training data set, testing data set is also needed to be normalized and de-noised. PCA model generated from training data and resampling are adopted to acquire signal patterns. The five signal patterns as a set represent current status of sampled raw data collected from the process in signal geometry manner, possessing major features for fault recognition.

After signal templates and signal patterns to be recognized are built, PC1DARMF algorithm can be performed step by step as demonstrated previously in Figure 4. The initial parameters for the algorithm are selected as follows: assigned value $m^{(0,j)}$, rank parameter $r^{(0)}$, and the threshold thm_M are subject to random choices while the initial length of filter mask $N^{(0)}$ is an integer chosen arbitrarily from 20 to 30 (0.3 to 0.5 times the trend length) for balancing tradeoff between algorithm effectiveness and low computational cost. The convergence rates α and β are pointed in Section 2.3. The algorithm stops when the variation of $e^{(itN)}$ is less than 1% within continuous 50 iterations and max iterations is set 200 to ensure algorithm convergence.

In this example, an unrecognized signal pattern of a selected PC adopts eight faulty statuses signal templates for that PC as supervisory signals and applies PC1DARMF algorithm to find its best matching signal template. If one signal pattern and its best matching signal template turn out to both derive from the same fault type, it is regarded as correct fault diagnosis, otherwise the incorrect diagnosis. Table 4 lists the correct diagnosis rates of 20 tests for each fault type using five PCs.

Table 4: Correct diagnosis rates of TEP deterministic faults using five PCs (20 simulations for each fault type)

PC i used:	PC1	PC2	PC3	PC4	PC5
Fault type					
IDV(0)	20%	20%	5%	0.0	10%
IDV(1)	60%	100%	0.0	0.0	90%
IDV(2)	65%	100%	75%	100%	15%
IDV(3)	60%	35%	50%	30%	10%
IDV(4)	15%	10%	0.0	10%	55%
IDV(5)	75%	0.0	0.0	5%	60%
IDV(6)	100%	85%	100%	90%	30%
IDV(7)	95%	40%	0.0	5%	90%

Li and Xiao EURASIP Journal on Advances in Signal Processing 2011 2011:83 doi:10.1186/1687-6180-2011-83

In order to combine the fault diagnosis results given by five PCs, multisource data fusion employing consensus theory is considered here to give a more reliable final result. The linear opinion pool (LIOP) as one of the most popular approaches of consensus theory achieves results fusion by computing the weighted sum of diagnosis credibility given by each PC.

$$C(w|PC) = \sum_i \lambda_i P(w_j|PC_i)$$

(26)

where $P(w_j|PC_i)$ quantifies the credibility of algorithm result of using ith PC for fault diagnosis. In other words, it reflects the frequencies of ascending orders of index FI_n (introduced in Figure 4) in overall FI values when choosing the right signal template of PC_j. $P(w_j|PC_i)$ can be calculated on the basis of statistics of extra training data set, which contains 10 simulation runs for each type with fault introduced time of the 8th h after the simulation started. λ_i is the weight of result given by PC_i and can be determined according to data variation captured by related PC. Tables 5 and 6 list $P(w_j|PC_i)$ and weight λ_i for PC_i. Based on the knowledge of Tables 5 and 6 and (26), the final diagnosis results using PC1DARMF algorithm and consensus theory for IDV(0) to IDV(7) are tabulated in Table 7.

Table 5: Credibility of PC1DARMF algorithm for deterministic fault diagnosis using PC_i

Rank	PC1	PC2	PC3	PC4	PC5
1	66.25%	55.00%	33.75%	41.25%	36.25%
2	16.25%	12.50%	23.75%	18.75%	25.00%
3	12.50%	21.25%	17.50%	13.75%	16.25%
4	5.00%	8.75%	8.75%	11.25%	10.00%
5	0.00	2.50%	13.75%	7.50%	6.25%
6	0.00	0.00	2.50%	6.25%	6.25%
7	0.00	0.00	0.00	1.25%	0.00
8	0.00	0.00	0.00	0.00	0.00

Li and Xiao EURASIP Journal on Advances in Signal Processing 2011 2011:83
doi:10.1186/1687-6180-2011-83

Table 6: Weight λ_i of PC_i for deterministic fault diagnosis

PCi	i
PC1	60.91%
PC2	15.63%
PC3	11.08%
PC4	7.43%
PC5	4.96%

Li and Xiao EURASIP Journal on Advances in Signal Processing 2011 2011:83
doi:10.1186/1687-6180-2011-83

Table 7: IDV(0) to IDV(7) diagnosis results using PC1DARMF algorithm and consensus theory

Fault type	Correct diagnosis rate
IDV(0)	20%
IDV(1)	60%
IDV(2)	65%
IDV(3)	60%
IDV(4)	15%
IDV(5)	75%
IDV(6)	100%
IDV(7)	95%

Li and Xiao EURASIP Journal on Advances in Signal Processing 2011 2011:83
doi:10.1186/1687-6180-2011-83

The fusion results of five PCs in Table 7 are the same with results only using signal templates of PC_1 in Table 4. It requires less computational effort to only use signal templates of PC_1 for deterministic faults diagnosis in TEP. Table 7 suggests normal operation (IDV(0)) is correctly diagnosed with low rate. When the process is under normal operation, observations are in steady state with minor oscillation in noise level. Then, the normalization and PCA dimension reduction may result in signal templates varying randomly rather than retaining regular geometry shapes.

Table 8 compares the performances of PC1DARMF algorithm with multivariate statistics based approaches (MSBA) [17]. Both MSBA and the proposed method are on-line monitoring methods. Besides, [17] is among handful investigations which gave the detailed specifications of data in use and studied all fault types of TEP. It helps provide more comprehensive comparisons. Twenty MSBA includes PCA, DPCA, FDA.DFDA, CVA, PLS, MS (multivariate statistics) based statistic measurement (such as Hotelling T^2 or Q statistic) [17] and average values of correct diagnosis rates of 20 MSBA are listed in Table 8. It shows four out of seven faults are more easily detected by PC1DARMF algorithm rather than MSBA. IDV(3) is defined as unobservable from the process data in [17], which implies no observable change in mean or the variance can be detected. All MSBA performs poorly on IDV(3) diagnosis. However, PC1DARMF algorithm manages to capture variations in signal geometries and performs much better than MSBA. IDV(4) only cause the mean and standard deviation of each variable differ less than 2% between the faulty status and normal operation (IDV(0)) [17]. This phenomenon leaves signal shapes of observations almost invariable from IDV(0), causing both misclassification rates for IDV(4) and IDV(0) higher than other faults. In general, both average values for seven deterministic faults are equally well. Considering the testing data for PC1DARMF algorithm are more diverse than MSBA in [17], the proposed method fares better than the existing ones.

Table 8: Comparisons of correct diagnosis rates for deterministic faults in TEP between PC1DARMF algorithm and MSBA

Fault type	PC1DARMF algorithm	MSBA [17]
IDV(1)	60%	88.76%
IDV(2)	65%	92.54%
IDV(3)	60%	12.41%
IDV(4)	15%	49.56%
IDV(5)	75%	69.44%
IDV(6)	100%	83.39%
IDV(7)	95%	75.33%
Average	67.14%	67.35%

Li and Xiao EURASIP Journal on Advances in Signal Processing 2011 2011:83
doi:10.1186/1687-6180-2011-83

Stochastic Fault Classification in TEP

IDV(8) to IDV(12) given in Table 3 are featured by random variations in measurements when one of them occurs. In this subsection, PC1DARMF algorithm is employed to classify stochastic fault types. The training data set consists of 10 simulation runs for each fault type and fault is introduced 8 h after simulation started. The testing data set simulates faulty states with different faults occurrence time and 20 simulation runs are provided for each fault type (with fault occurrence time 4th, 2nd, 10th, 6th h per five simulations). Parallel analysis suggests that five PCs capture most variations. The initial parameter settings for algorithm follow the settings in Section 4.2. Five sets of signal templates for characterizing IDV(8) to IDV(12) are depicted in Figure 8. With the same steps in Section 4.2, Tables 9, 10, 11 and 12 list corresponding statistics for stochastic fault classification directly for saving details of derivations.

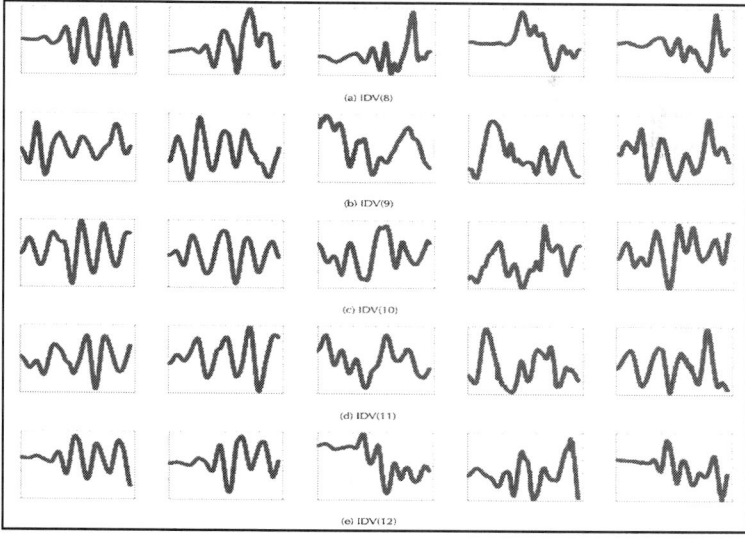

Figure 8: Signal templates for (a) IDV(8) to (e) IDV(12). In each row, five figures from left to right correspond to signal templates of PC_1 to PC_5, respectively.

Table 9: Correct classification rates of TEP stochastic faults using signal templates of five PCs (20 simulations for each fault type)

PCi used:	PC1	PC2	PC3	PC4	PC5
Fault type					
IDV(8)	20%	25%	0.0	0.0	0.0
IDV(9)	40%	75%	5%	5%	45%
IDV(10)	0.0	0.0	65%	70%	5%
IDV(11)	60%	25%	30%	35%	45%
IDV(12)	20%	35%	30%	70%	0.0

Li and Xiao EURASIP Journal on Advances in Signal Processing 2011 2011:83 doi:10.1186/1687-6180-2011-83

Table 10: Credibility of PC1DARMF algorithm for stochastic fault classification using PC_i

Rank	PC1	PC2	PC3	PC4	PC5
1	34.00%	32.00%	26.00%	48.00%	36.00%
2	36.00%	22.00%	40.00%	22.00%	16.00%
3	14.00%	22.00%	18.00%	12.00%	14.00%
4	10.00%	10.00%	8.00%	12.00%	20.00%
5	6.00%	14.00%	8.00%	6.00%	14.00%

Li and Xiao EURASIP Journal on Advances in Signal Processing 2011 2011:83 doi:10.1186/1687-6180-2011-83

Table 11: Weight λ_i of PC_i for stochastic fault classification

PCi	i
PC1	51.18%
PC2	29.70%
PC3	7.78%
PC4	5.94%
PC5	5.40%

Li and Xiao EURASIP Journal on Advances in Signal Processing 2011 2011:83
doi:10.1186/1687-6180-2011-83

Table 12: IDV(8) to IDV(12) classification results using PC1DARMF algorithm and consensus theory

Fault type	Correct diagnosis rate
IDV(8)	20%
IDV(9)	35%
IDV(10)	5%
IDV(11)	45%
IDV(12)	45%

Li and Xiao EURASIP Journal on Advances in Signal Processing 2011 2011:83
doi:10.1186/1687-6180-2011-83

In Table 9 statistics suggest that supervision of signal template sets of every PC results in at least one 0 correct classification rate, while the fusion gives more comprehensive results. The performances of stochastic faults classification are poorer than performances of deterministic faults. One important reason was analyzed in Section 4.2 for random variations hinder formation of standard patterns. Table 13 tabulates comparisons between PC1DARMF algorithm and MSBA[17] for classifying stochastic faults in TEP. The former performs better on two out of five faults. Note that IDV(9) is also viewed as unobservable like IDV(3) in [17] and very difficult to be identified by MSBA. PC1DARMF algorithm here again fares better for unobservable fault. Since random variations lead to irregular morphology of signal shape but observable quantitative variations of statistic measurements, PC1DARMF algorithm performs poorer than MSBA on average. However, the conclusion that MSBA will always give lower misclassification rates than PC1DARMF algorithm would be incorrect, because 8 out of 20 approaches studied in [17] still give lower average correct diagnosis rates compared to PC1DARMF algorithm. If more diverse training data types are provided, the classification results of the proposed method are expected to be better.

Table 13: Comparisons of correct diagnosis rate for stochastic faults in TEP between PC1DARMF algorithm and MSBA

Fault type	PC1DARMF algorithm	MSBA [17]
IDV(8)	20%	60.29%
IDV(9)	35%	9.52%
IDV(10)	5%	47.41%
IDV(11)	45%	36.51%
IDV(12)	45%	67.44%
Average	30%	44.23%

Li and Xiao EURASIP Journal on Advances in Signal Processing 2011 2011:83 doi:10.1186/1687-6180-2011-83

Diagnosis of All Fault Types in TEP

After investigating diagnostic performances of PC1DARMF algorithm for two major fault classes, respectively, we proceed to study the case when it is applied to all possible faults in TEP. The training data set consists of 10 simulation runs for each fault type and the fault is introduced the 8th h after simulation started. The testing data set simulates faulty states with different faults occurrence time and 20 simulation runs are provided for each type (with fault introduced time 4th, 2nd, 10th, 6th h per five simulations). Parallel analysis is applied to find that six PCs enough capture most variations. The parameter selection rules for PC1DARMF algorithm is the same as previous. Tables 14, 15, 16 and 17 list related statistics. Table 17 suggests that the performances of PC1ADRMF algorithm degrades when more sets of signal templates representing more fault types are provided in comparisons with Tables 7, 8, 9, 10, 11 and 12. This phenomenon implies as a supervised pattern matching strategy, PC1DARMF algorithm may require more different training data covering major features of relevant groups as much as possible to assist forming highly representative signal templates.

Table 14: Correct diagnosis rates of IDV(0) to IDV(20) using signal templates of six PCs

PCi used:	PC1	PC2	PC3	PC4	PC5	PC6
Fault type						
IDV(0)	5%	60%	0.0	55%	0.0	10%
IDV(1)	0.0	0.0	50%	0.0	95%	5%
IDV(2)	95%	10%	100%	100%	90%	0.0
IDV(3)	0.0	0.0	0.0	5%	15%	0.0
IDV(4)	5%	0.0	0.0	5%	0.0	0.0
IDV(5)	25%	0.0	10%	0.0	5%	45%
IDV(6)	75%	100%	100%	80%	30%	5%
IDV(7)	55%	0.0	20%	0.0	100%	85%
IDV(8)	0.0	0.0	0.0	5%	0.0	0.0
IDV(9)	0.0	20%	0.0	10%	0.0	10%
IDV(10)	0.0	0.0	0.0	0.0	5%	0.0
IDV(11)	20%	5%	0.0	5%	25%	0.0
IDV(12)	5%	0.0	0.0	0.0	0.0	55%
IDV(13)	5%	0.0	5%	5%	20%	15%
IDV(14)	5%	0.0	0.0	5%	0.0	0.0
IDV(15)	10%	5%	0.0	5%	5%	10%
IDV(16)	0.0	0.0	0.0	0.0	0.0	5%
IDV(17)	65%	50%	20%	35%	5%	10%
IDV(18)	25%	35%	65%	50%	30%	20%
IDV(19)	0.0	0.0	10%	0.0	0.0	0.0
IDV(20)	0.0	0.0	5%	0.0	10%	0.0

Li and Xiao EURASIP Journal on Advances in Signal Processing 2011 2011:83 doi:10.1186/1687-6180-2011-83

Table 15: Credibility of PC1 DARMF algorithm for IDV(0) to IDV(20) diagnosis using PC_i

Rank	PC1	PC2	PC3	PC4	PC5	PC6
1	19.05%	17.14%	22.86%	20.48%	26.19%	20.48%
2	11.90%	6.67%	12.86%	11.43%	6.19%	8.57%

3	9.05%	5.71%	10.00%	10.00%	5.24%	4.29%
4	9.52%	7.62%	8.57%	4.76%	4.76%	4.76%
5	4.76%	5.71%	3.81%	4.76%	7.14%	7.26%
6	6.19%	7.62%	5.24%	7.14%	5.71%	5.71%
7	7.62%	8.10%	3.81%	7.14%	5.24%	6.67%
8	5.71%	5.71%	3.33%	5.71%	5.24%	4.29%
9	2.38%	7.62%	2.38%	2.86%	2.86%	2.86%
10	6.67%	5.24%	3.81%	3.81%	7.14%	4.29%
11	3.33%	5.24%	4.29%	6.19%	9.52%	1.43%
12	6.67%	5.24%	5.24%	5.24%	3.33%	4.76%
13	2.86%	1.90%	2.86%	3.81%	0.95%	2.38%
14	2.38%	5.24%	3.81%	0.48%	2.38%	3.81%
15	0.48%	0.48%	2.38%	0.95%	0.95%	2.86%
16	0.48%	1.43%	1.90%	1.90%	3.33%	3.81%
17	0.48%	1.43%	0.48%	1.90%	0.95%	5.24%
18	0.0	0.95%	0.48%	1.43%	1.90%	3.81%
19	0.0	0.0	1.43%	0.0	0.95%	1.43%
20	0.0	0.48%	0.48%	0.0	0.0	0.95%
21	0.48%	0.48%	0.0	0.0	0.0	0.0

Li and Xiao EURASIP Journal on Advances in Signal Processing 2011 2011:83
doi:10.1186/1687-6180-2011-83

Table 16: Weight λ_i of PC_i for IDV(0) to IDV(20) diagnosis

PC_i	i
PC1	41.07%
PC2	27.48%
PC3	13.03%
PC4	9.60%
PC5	4.60%
PC6	4.23%

Li and Xiao EURASIP Journal on Advances in Signal Processing 2011 2011:83
doi:10.1186/1687-6180-2011-83

Table 17: IDV(0) to IDV(20) diagnosis results using PC1DARMF algorithm

Fault type	Correct diagnosis rate
IDV(0)	15%
IDV(1)	30%
IDV(2)	95%
IDV(3)	0.0
IDV(4)	5%
IDV(5)	25%
IDV(6)	100%
IDV(7)	65%
IDV(8)	0.0
IDV(9)	5%
IDV(10)	0.0
IDV(11)	15%
IDV(12)	0.0
IDV(13)	5%
IDV(14)	5%
IDV(15)	5%
IDV(16)	0.0
IDV(17)	85%
IDV(18)	30%
IDV(19)	0.0
IDV(20)	0.0

Li and Xiao EURASIP Journal on Advances in Signal Processing 2011 2011:83
doi:10.1186/1687-6180-2011-83

CONCLUSIONS AND DISCUSSION

In this article, a supervised pattern classification method using one-dimensional adaptive rank-order morphological filter called PC1DARMF is developed to detect and recognize different faults in Tennessee Eastman process. This method generates several signals of featured geometry shapes as standard patterns on the basis of training data. With the same processing procedures as training data, testing data reflecting current operational states of TEP are transformed to signal patterns with defined specification. They are matched against standard signal patterns with employment of 1DARMF. It adaptively adjusts filter mask and rank parameter for each sample of signal rather than adopting uniform ones for all the samples. The major parameters for implementing this algorithm are capable of being randomly chosen.

TEP deterministic, stochastic, and all fault classes diagnosis are studied to verify the effectiveness of the proposed method in complex process. Consensus theory is employed for fusion of results provided by different sources The results show with only small quantity of training data provided and the testing data being much more different from training data, the performances of deterministic or stochastic faults diagnosis is better than or equally well as multivariate statistics based approaches studied in [17]. Several faults deemed as unobservable for multivariate statistic based approaches can be also recognized more easily. Deterministic faults diagnosis fares better than the stochastic faults diagnosis, since deterministic ones are apt to form similar or regular trends regardless of specified faulty conditions and noise levels while stochastic ones fail to retain basic morphologies for signal patterns. It is also noted that for some real-time applications, signal template sets provided by only one or two PCs is recommended to reduce computational cost. The future work also lies in diagnosing more diverse or multiple faults in TEP or other complex processes to improve the proposed methods. Besides, Ku et al. [19] pointed DPCA model was expected to perform better than regular PCA on the TEP problem. DPCA could be introduced to extract more information in data set.

Despite some promising results of the proposed method, this article presents only a preliminary implementation in complex process, which sheds light on the novel idea of PC1DARMF. The key concept of this idea is to achieve pattern matching between standard pattern and unrecognized pattern by 1DARMF. The specification of pattern defined is not only limited to time-domain signal geometry shapes or even not signal geometry shapes. Frequency spectrums, power spectrum, vector bases, and feature coefficients, etc. can be assembled to form user-defined pattern as well. The new developed pattern should be capable of not only capturing different characteristics for each group but also maintaining a relatively steady form without too much unexpected random variations. Using the same scheme may address the problems such as unstable patterns introduced by random variations.

ACKNOWLEDGEMENTS

This work was supported by National Natural Science Foundation of China No. 60736026 and No. 60904044 grants and the authors would like to thank the control scheme code for TEP fault diagnosis provided by Evan L. Russell, Leo H. Chiang and Richard D. Braatz, Large Scale Systems Research Laboratory, Department of Chemical Engineering, University of Illinois at Urbana-Champaign. The authors also would like to appreciate the anonymous reviewers whose comments greatly enhanced the presentation and clarity of this paper.

REFERENCES

1.   M Iri, K Aoki, E O'Shima, H Matsuyama, An algorithm for diagnosis of system failures in the chemical process. Comput Chem Eng 3(1-4), 489–493 (1979).

2.   MA Paulonis, JW Cox, A practical approach for large-scale controller performance assessment, diagnosis, and improvement. J Process Control 13(2), 155–168 (2003).

3.   V Venkatasubramanian, R Rengaswamy, K Yin, SN Kavuri, A review of process fault detection and diagnosis part I: quantitative model-based methods. Comput Chem Eng27(3), 293–311 (2003).

4.   V Venkatasubramanian, R Rengaswamy, SN Kavuri, A review of process fault detection and diagnosis part II: qualitative models and search strategies. Comput Chem Eng 27(3), 313–326 (2003).

5.   V Venkatasubramanian, R Rengaswamy, SN Kavuri, K Yin, A review of process fault detection and diagnosis part III: process history based methods. Comput Chem Eng 27(3), 327–346 (2003).

6.   R Stevenson, G Arce, Morphological filters Statistics and further syntactic properties. IEEE Trans Circuits Syst 34(11), 1292–1305 (1987).

7.   I Pitas, AN Venetsanopoulos, Nonlinear Digital Filters: Principles and Applications (Kluwer Academic Publishers, Boston, 1990), p. 1

8.   P Salembier, Adaptive rank order based filters. Signal Process 27(1), 1–25 (1992).

9.   P Salembier, M Kunt, Multiresolution decomposition and adaptive filtering with rank order based filters--application to defect detection. Proceedings of IEEE International Conference Acoustics, Speech and Signal Process (Toronto, Canada, 1991), pp. 2389–2392

10. A Feuer, E Weinstein, Convergence analysis of LMS filters with uncorrelated Gaussian data. IEEE Trans Acoust Speech Signal Process 33(1), 222–230 (1985).

11. NG Nikolaou, IA Antoniadis, Application of morphological operators as envelope extractors for impulsive-type periodic signals. Mech Syst Signal Process 17(6), 1147–1162 (2003).

12. JJ Downs, EF Vogel, A plant-wide industrial process control problem. Comput Chem Eng17(3), 245–255 (1993)

13. PR Lyman, C Georgakis, Plant-wide control of the Tennessee Eastman problem. Comput Chem Eng 19(3), 321–331 (1995).

14. A Raich, A Cinar, Multivariate statistical methods for monitoring continuous processes: assessment of discrimination power of disturbance models and diagnosis of multiple disturbances. Chemomet Intel Lab Syst 30(1), 37–48 (1995).

15. A Raich, A Cinar, Statistical process monitoring and disturbance diagnosis in multivariate continuous processes. AIChE J 42(4), 995–1009 (1996).

16. A Raich, A Cinar, Diagnosis of process disturbances by statistical distance and angle measures. Comput Chem Eng 21(6), 661–673 (1997).

17. EL Russell, LH Chiang, RD Braatz, Data-driven methods for fault detection and diagnosis in chemical processes. Springer, New York, 13–162 (2000)

18. LH Chiang, EL Russell, RD Braatz, Fault diagnosis in chemical processes using Fisher discriminant analysis, discriminant partial least squares, and principal component analysis. Chemomet Intel Lab Syst 50(2), 243–252 (2000).

19. W Ku, RH Storer, C Georgakis, Disturbance detection and isolation by dynamic principal component analysis. Chemomet Intel Lab Syst 30(1), 179–196 (1995).

20. M Kano, K Nagao, S Hasebe, I Hashimoto, H Ohno, R Strauss, BR Bakshi, Comparison of statistical process monitoring methods: application to the Tennessee Eastman challenge problem. Comput Chem Eng 26(2), 161–174 (2002).

21. J-M Lee, S Joe Qin, Fault detection and diagnosis based on modified independent component analysis. AIChE 52(10), 3501–3514 (2006).

22. J Chen, C-M Liao, Dynamic process fault monitoring based on neural network and PCA. J Process 12(2), 277–289 (2002).

23. MN Nashalji, MA Shoorehdeli, M Teshnehlab, Fault detection of the Tennessee Eastman process using improved PCA and neural classifier. Soft computing in industrial applications75, 41–50 (2010).

24. LH Chiang, ME Kotanchek, AK Kordon, Fault diagnosis based on Fisher discriminant analysis and support vector machines. Comput Chem Eng 28(8), 1389–1401 (2004)

25. A Kulkarni, VK Jayaraman, BD Kulkarni, Knowledge incorporated support vector machines to detect faults in Tennessee Eastman process. Comput Chem Eng 29(10), 2128–2133

26. R Eslamloueyan, Designing a hierarchical neural network based on fuzzy clustering for fault diagnosis of the Tennessee-Eastman process. Appl Soft Comput 11(1), 1407–1415 (2011).

27. S Verron, T Tiplica, A Kobi, Distance rejection in a bayesian network for fault diagnosis of industrial systems. Proceedings of 16th Mediterranean Conference on Control and Automation (Ajaccio, France, 2008), pp. 615–620

28. S Verron, T Tiplica, A Kobi, Fault diagnosis with bayesian networks: application to the Tennessee Eastman Process. Proceedings of 2006 IEEE International Conference on Industrial Technology, 98–103 (2006)

29. MR Maurya, R Rengaswamy, V Venkatasubramanian, Application of signed digraphs-based analysis for fault diagnosis of chemical process flowsheets. Eng Appl Artif Intell 17(5), 501–518 (2004).

30. MR Maurya, R Rengaswamy, V Venkatasubramanian, Fault diagnosis by qualitative trend analysis of the principal components. Chem Eng Res Des 83(9), 1122–1132 (2005).

31. MR Maurya, R Rengaswamy, V Venkatasubramanian, Fault diagnosis using dynamic trend analysis: a review and recent developments. Eng Appl Artif Intell 20(2), 133–146 (2007).

32. F Akbaryan, PR Bishnoi, Fault diagnosis of multivariate systems using pattern recognition and multisensor data analysis technique. Comput Chem Eng 25(9-10), 1313–1339 (2001).

33. A Singhal, DE Seborg, Evaluation of a pattern matching method for the Tennessee Eastman challenge process. J Process Control 16(6), 601–613 (2006).

34. G Lee, C Han, ES Yoon, Multiple-fault diagnosis of the Tennessee Eastman Process based on system decomposition and dynamic PLS. Ind Eng Chem Res 43, 8037–8048 (2004).

35. N Lu, F Wang, F Gao, Combination method of principal component and wavelet analysis for multivariate process monitoring and fault diagnosis. Ind Eng Chem Res 42, , 4198–4207 (2003)

36. DL Donoho, De-nosing by soft-thresholding. IEEE Trans Inf Theory 41(3), 613–627 (1995).

37. JA Benediktsson, PH Swain, Consensus theoretic classification methods. IEEE Trans Syst Man Cybern 22(4), 688–704 (1995)

38. J Edward Jackson, A User's Guide to Principal Components (Wiley, New York, 2003), pp. 46–47

The Influence of Engine Fuel Manufacturing Processes on Their Performance Properties in Operating Conditions

Krzysztof Biernat[1, 2]

[1]Department of Fuels, Biofuels and Lubricants, Automotive Industry Institute, Poland
[2]Institute for Ecology and Bioethics of CSWU, Poland

INTRODUCTION

Straight-Run Processing of Petroleum

Straight-run processing of petroleum (or crude oil) is the kind of processing where the material is distilled after being purified and is separated into fractions (or cuts), based on boiling range differences between its respective components, without modifying the chemical structure of its constituent compounds. Fractional distillation is typically carried out in two steps: under normal pressure and under reduced pressure. Figure 1 shows a simplified diagram for straight-run processing of petroleum.

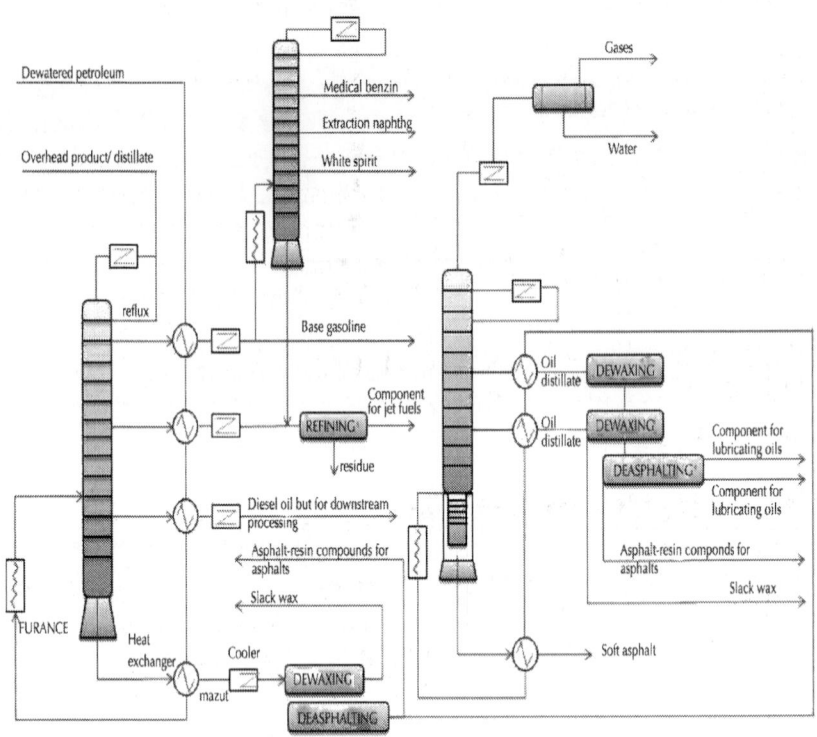

Figure 1: Straight-run processing of petroleum – simplified diagram.

Before processing, petroleum has a content of solid contaminants (up to 1.5 %) and water (up to 0.3 %). Part of insoluble contaminants and water are removed from petroleum by being left to stand (sedimentation). The process simply involves storage of petroleum in storage tanks for a certain length of time, during which the solids and part of its water content migrate to the bottom while the other solids dissolve in the water, forming a brine, which it is rather hard to remove. In that emulsion, the petroleum is the dispersion medium while water with salts is the disperse phase. Existing emulsion breaking methods are categorized into three groups:

- mechanical methods – involve sedimentation, centrifugation, and filtration of fresh emulsions;
- chemical methods – involving first of all the use of deemulsifiers, which are supposed to dissolve the adsorption film at the water interface;
- electrical methods – involving the use of electrohydrators.

In all such groups, deemulsification involves the combining of small disperse water drops by overcoming surface tensions forces at the interface, to form drops big enough to be able to fall down the tank by gravitation. In industry, electrical methods are typically used, where the reactor has two flat electrodes installed in it, between which there is a voltage of about 30 kV. The water molecules move towards the respective electrodes and, as the electrode symbol changes, the molecules lose their electrical orientation and collide with one another, forming suitably large drops which fall down the tank. The resulting dewatered petroleum is taken at the top of the reactor. Electrohydration is carried out at an elevated temperature in a continuous manner.

The purified and degassed petroleum is made to flow into a tubular furnace to be heated to a temperature not higher than 370°C and then to the base of a first distillation tower (atmospheric column). The distillation tower is approximately 40 m high and comprises a number of so-called trays. Their shape depends on the rectification process parameters, the typical tray designs include: bubble cap, sieve, wave, West, cascade, Venturi, ejector, valve, and combinations of the above. The rectification process in the atmospheric tower is carried out at nearly-atmospheric pressures. After flowing into the tower, the hot petroleum is divided into fractions as follows: the consecutive fractions (in terms of density) evaporate and then condense on the respective tray groups, are then taken to heat exchangers (where their heat is transferred to the petroleum flowing into the furnace) and sent downstream.

Inside the column, there circulates a so-called "reflux", which may be of the hot (water or, less frequently, feedstock), cold, or circulating type. The reflux is supposed to keep sufficient temperatures, especially in the top regions of the column.

The products obtained in the atmospheric distillation tower include:

- overhead fraction (which also functions as the circulating reflux;

- side fractions (their amount and boiling ranges depend on the process parameters and design of column); typically, they include:

  light gasoline fraction;

  naphtha fraction;

  kerosine fraction (jet fuels);

  diesel fuel fraction;

  spindle oil fraction.

- distillation residue, also called mazut.

For a better fractionation of the gasoline fraction (to obtain medical benzin, extraction naphtha, painter's naphtha), it is possible to re-rectify a portion on it in a separate atmospheric tower. Its design and processing regime are the same as described earlier. Distillation residue from that column is combined with the kerosine fraction. Rectification in each atmospheric tower takes place above the feed point for the gas phase but below the feed point for the liquid phase. In order to ensure appropriate rectification of the liquid phase, it is necessary to provide additional heat or another evaporating agent. Rectification in atmospheric towers is a continuous process.

A vacuum tower is another major component of a straight-run petroleum processing plant. It serves for distillation of the high-boiling distillation residue from the atmospheric tower, after heating it to 430°C in deep-vacuum conditions. In the vacuum towers of to-day, pressures in the evaporation region are in the range (1.99...2.27) kPa, compared with 0.66 kPa at the vapor outlet point. Atmospheric towers are 15-30 m high, their diameter is a max. of 12 m. The column is filled with bubble cap or sieve or valve-type trays; depending on the kind of product made in the vacuum column, the number of trays varies between 8...14 (if the product is used for catalytic cracking) and 38...42 (if the distillate is used for oil production).

Mazut, which flows into the column typically above tray 6, is combined with foam reducers (for instance, silicones at a concentration of 0.75 mg/dm^3 of feedstock). Lower pressure, which is reduced by means of a system of so-called ejectors and condensers, leads to lower boiling ranges, enabling rectification of heavy mazut without decomposing it.

The vacuum tower products include: between one and three side fractions, a gas overhead fraction and the distillation residue called soft asphalt. After further straight-run or destructive processing, the side fractions are an important component for various engine oils and heating oils.

In addition, straight-run processing comprises a number of auxiliary treatments, enabling the various components to be separated from the distillation products. Such treatments include: crystallization and filtration, refining with the use of H_2SO_4 and selective solvent, dewaxing with the use of selective solvents and de-asphalting with propane.

Crystallization and filtration are intended to separate paraffins from the distillation cuts. The crystallized paraffins are then filtered off using a pressure press and filtration cloth. After the removal of paraffins, the oil fraction is further processed and the crude paraffin that results is used for making insulation materials, maintenance materials, or candles (after refining). Refining of the post-distillation fractions is intended to remove asphalts and resinous compounds, which are undesirable in further processing.

Chemical Refining

In order to remove acid oxygen-based compounds or some sulfur compounds, the petroleum products are subjected to refining by means of lye: it reacts with acidic compounds, forming respective water-soluble salts. Part of the compounds remain in the refined product and can only be removed therefrom by washing with water. The oils are refined using weak sodium hydroxide solutions. The process is carried out at elevated temperatures to prevent the formation of emulsions which it would otherwise be hard to break.

In the process of neutralization of the oil distillates with lye, the napthene acid content of the distillate initially reacts with sodium hydroxide to form soaps, which dissolve in aqueous lye solutions whereby they are removed from the distillate; the emulsions formed then are usually not very stable. Stable emulsions are formed after another water-washing operation. This is due to the presence in the oil of resinous products which, in a dispersed form, are hydrophobic emulsifiers. Owing to the presence of high levels of hydrophilic emulsifiers, their activity as hydrophilic emulsifiers is not manifested during neutralization; on the other hand, when washing with water, the hydrophobic emulsifiers are not separated from the oil along with the waste lye and their effect shows very well. The phenomenon is observed when refining oils having a high content of tars and asphalts and a high content of oxidation products. The hydrophobic emulsions formed then may be separated by heating the distillates to high temperatures, by adding a solution of hydrophilic naphthene soaps, or by treatment with the use of weak solutions of mineral acids, which destroy the emulsifier's interface films.

Refining With the Use of Selective Solvents

Refining the kerosine fractions with solvents is based on the choice of a suitable solvent which is able to dissolve in different ways the desirable and the undesirable components of the refined product. Selective solvents are expected to be able to readily dissolve, to be selective and stable, to form readily separating extract and raffinate phases, to be easily regenerated, to resists corrosion in operating conditions, and to have non-toxic properties.

The essential parameters which determine the level of refining include temperature and the solvent-to-material ratio. The choice of temperature for the refining process depends on the critical temperature of solubility for a given mixture. Refining with the use of selective solvents is feasible in those temperature ranges where a two-phase system exists: the raffinate solution (containing a trace amount of solvent) and the extract solution (comprising mainly the solvent and the undesirable components of the starting raw material which are dissolved in the solvent). The critical solubility temperature depends on the structure of hydrocarbon molecules: their critical solubility temperature is lower (and going down rapidly) for higher numbers of rings in the hydrocarbons but is lower for longer alkyl lengths. Naphthenes with five-member molecules will better reduce the critical temperature of solubility, compared with six-member molecules. In the case of aromatic hydrocarbons and naphthenes with same structures, for same solvent, critical temperature of solubility of aromatic hydrocarbons is much lower than that of naphthenes. Naphthene-aromatic hydrocarbons have lower critical temperatures of solubility, compared with naphthene hydrocarbons having a similar structure. Normal paraffins have the highest critical temperature of solubility. The value of critical temperature of solubility and kerosine fractions in a solvent is affected also by the solvent's properties: for instance, critical temperature of solubility of hydrocarbons in nitrobenzene is much lower than that in phenol, but is lower in phenol compared with that in furfurol.

Solubility of substances in solvents depends on attractive forces between the molecules of the solvent and the solute. Attraction between molecules takes place due to the Van der Waals forces and hydrogen bonds. In view of the fact that kerosine fractions comprise mainly nonpolar hydrocarbons, selective extraction of undesirable components is only possible in the case of Deby's effect, that is, co-operation of induced dipols which are formed in non-polar molecules under the effect of the electric field of polar solvent's molecules. The highest polarizability is shown by aromatic hydrocarbons, the lowest by naphthenes and paraffins. Therefore, aromatic hydrocarbons readily submit to the action of the electric field of solvents, which leads to the formation in their molecules of an induced dipol moment, resulting in their readily dissolving in polar solvents.

In addition to the refining temperature and type of solvent, the degree of extraction of undesirable components depends also on its amount that is indispensable for extraction. On the other hand, the amount of solvent depends on its properties, chemical composition of the starting material, the desirable refining degree, and on the extraction method.

Selective refining by means of furfurol is a method for removing aromatic hydrocarbons from vacuum petroleum distillates, and is used as a base oil production step. Furfurol is a polar substance with a high dipol moment. It is able to selectively dissolve hydrocarbons by inducing the dipol moment in the hydrocarbon molecules which are contacted with furfurol. It is useful as a solvent in selective refining processes because of the following advantages:

- high density and lack of tendency to form foams or emulsions; these properties facilitates phase separation between the raffinate extract solutions;
- low freeze point: therefore, its mixtures are easier to handle at low temperatures, requiring no extra care or devices;
- large difference between the critical temperatures of solubility for paraffinic and aromatic compounds.

On the other hand, furfural has the following disadvantages:

- low resistance to oxidation at high temperatures, both in alkaline and acidic environments,
- formation of acidic oxidation products and high-molecular products of polycondensation;
- high toxicity.

In the process of selective refining with furfurol, aromatic compounds are removed from the oil more readily than paraffins, compounds with high viscosity. Hence, a more aromatic compound requires less solvent and lower temperatures to be entirely dissolved. Therefore, by selecting suitable extraction temperatures and solvent-to-material ratio, it is possible to remove either only aromatic compounds from the raw material or – after modification of extraction conditions – to remove mixed compounds as well.

Refining by Adsorption

Adsorption as a refining process is currently used, first of all, in the finishing of light kerosine cuts, lubricating oils, specialty oils, and paraffins.

The role of adsorption in the refining of petroleum products is in the adsorption of asphaltenes, resins, diolefins, acids, etc. on the adsorbent surface, consequently providing a finished product with improved color and odor, and stable physico-chemical and performance properties.

The adsorption refining process is carried out either by the cold or hot method, using percolation (where adsorbent pellets are used) or by the contact method, using so-called decolorizing earths (adsorbents in the pulverized form, obtained from natural aluminosilicates).

The following materials are used in the refining process:

- sorbents, obtained by thermal or thermal-chemical modification of natural mineral raw materials (aluminosilicates);
- synthetic sorbents, such as: silica gel or alumina;
- active carbon.

When selecting s suitable sorbent, care is taken not only about the efficiency of regeneration, connected with improving the properties of oil, but also about the cost-efficiency of the process. To select the most suitable sorbent, it is necessary to consider some of its properties, first of all, its refining capacity, selectivity, chemical properties, mechanical strength, costs, availability, possible reactivation, and disposal.

DESTRUCTIVE PROCESSING OF PETROLEUM

Destructive processing of petroleum involves modification of the structure of hydrocarbons contained in the fractions obtained from the straight-run processing of petroleum. Such modification is intended to improve intermediates for use in final product blending. Destructive processing cannot be carried out with the omission of straight-run processing. The essential process groups included in destructive processing of petroleum are discussed below.

Thermal Cracking

Thermal cracking is a process in which large hydrocarbon molecules are broken to form light unsaturated hydrocarbons in high-temperature conditions.

Thermal cracking comprises three groups of processes:

- Cracking of liquid raw materials at high pressures (1961.3...6864.6) kPa in the temperature range (470...540) °C to obtain gasoline. The process is intended to obtain a higher amount of fuels at the cost of oil fractions. Gasoline can be obtained from the post-distillation side-fractions in atmospheric and vacuum towers, while heating oils can be obtained from the distillation residue (soft asphalt) in a vacuum tower. The process to obtain heating oils, carried out in mild conditions, is called *visbreaking*;

- Low-pressure cracking – also called coking, or destructive distillation. The process is carried out at temperatures in the range (450...550)°C. It is intended to provide light-colored products with a high hydrogen content, such as gasoline, diesel fuels or gases, as a result of decarbonization (concentrating asphalts and resins into so-called "petroleum coke"). The coke product is often a target product, intended for making, for instance, coatings for electrodes. The coking process is carried out as shown in the diagram below and may be interrupted or slowed down at any time by injection of an extra amount of cold raw material.

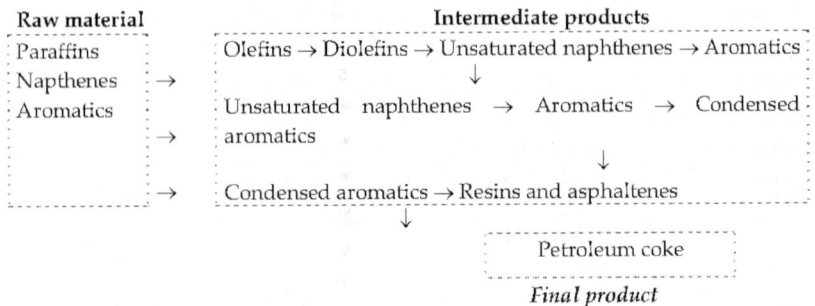

Raw material	Intermediate products

Thermal cracking in the most severe conditions: pyrolysis. The process is carried out at temperatures in the range (670...800)°C (though the process temperature may be as high as 1200°C). The process is intended mainly to provide unsaturated gases, usually ethylene, for use in petrochemical syntheses. The process also provides aromatic hydrocarbons such as benzene, toluene, xylenes, or naphthalene, and so-called post-pyrolytic gasoline which is a component for automotive gasoline, though they are only to be considered as side products.

In addition to the above, there exist a number of intermediate thermal cracking processes, for instance, vapor phase cracking in low pressure conditions at a temperature of 600°C to produce gasoline, or coking of the residue in severe conditions in order to increase the amount of gas and aromatization of liquid products.

Catalytic Cracking

Catalytic or thermocatalytic cracking processes are carried out at high temperatures in the presence of catalysts. They are intended to provide light products with good quality at the cost of heavy products (mainly gasoline and diesel fuels) or to improve the quality of other distillation products.

Gasoline and diesel fuels, obtained at temperatures in the range (450...500)°C in the presence of an aluminosilicate catalyst, are characterized by high resistance to decomposition and oxidation processes (mainly gasoline) and high purity. The mechanism of catalytic cracking is reverse to that of thermal cracking and leads to highly saturated hydrocarbons.

Gasoline reforming has been isolated from the catalytic cracking and is a separate process, intended to improve the gasoline fractions by their aromatization and purification to remove sulfur compounds therefrom. Pure aromatic hydrocarbons such as benzene, toluene, xylenes, can be obtained from the aromatized gasoline for petrochemical synthesis after its suitable separation. Moreover, reforming provides hydrogen for hydrogen processes. Depending on its variant, the reforming process uses a number of catalysts (Co, Ni, Mo, Pt, Fe), usually aluminosilicates. The process temperature is around 550°C. The essential reactions taking place during gasoline reforming include dehydrogenation of cycloalkanes (naphthenes), dehydroisomerization of naphthenes, and dehydrocyclization of alkanes (paraffins). These reactions are accompanied by isomerization and hydrocracking of paraffins. The essential reactions generate free hydrogen, therefore, such reactions as desulfurization and saturation of alkenes take place as well.

Fluidized-bed Catalytic Cracking

Fluidized-Bed Catalytic Cracking (FBCC) of de-asphalted vacuum and heavy petroleum cuts, in the presence of aluminosilicate catalysts (typically zeolites), is one of the major methods for deep processing of petroleum that are used in advanced refineries. The process is highly complex in terms of equipment and, accordingly, involves relatively high investment costs. On the other hand, the use of the process unit is justified in economic terms, since on average, only about 50 % (m/m) of petroleum is distilled-off at an atmospheric pressure. The petroleum fraction that results from vacuum distillation, having a boiling range of (350...500)°C and constituting typically 25 % of its weight, is a perfect feedstock for the FBCC plant, for making valuable components of engine fuels and light olefins for use in synthetic plastics (polyethylene, polypropylene, rubbers, etc).

Cracking of high-molecular hydrocarbons causes breaking of intermolecular bonds, which is accompanied by dehydrogenation and hydrogenation, comprising hydrogen transfer reactions. The bonds between carbon atoms are broken in irreversible reactions. Out of a great variety of bonds between the atoms, those with the lowest energy are the easiest to break. The elementary energy of C-C bonds in paraffins is 265 kJ/mol, for C-H bonds it is 360 kJ/mol, and that of C-C bonds in aromatic compounds is (500...610) kJ/mol, therefore, paraffins are most frequently subjected to cracking. Hydrogen transfer reactions contribute to the formation of gasoline compounds as saturated compounds, though at the cost of formation of those with a low hydrogen content, including coke. During the cracking process, owing to thermodynamic conditions, polymerization of olefins – though only insignificant – is the first phase in the formation of aromatic compounds and coke.

The essential reaction leading to the formation of coke is the condensation of aromatic hydrocarbons with olefins. Therefore, naphthenes and naphthene-paraffin compounds are the most preferable raw materials for fluidized-bed cracking. On the other hand, aromatic feedstock hinders the cracking process, favoring the formation of coke.

During the cracking process, primary reactions are accompanied by a number of secondary ones which become more and more intensified: such processes include polymerization, aromatization, isomerization, alkylation and dealkylation. Catalytic cracking takes place at lower temperatures, compared with thermal cracking but the amount of coke being formed in it is much more limited. The aluminosilicate catalysts used in the process accelerate the most desirable reactions: the rate of cracking of paraffins is 10 times as high, compared with that in a purely thermal process, conversion of naphthenes is 1000 times as fast, and that of side-chain aromatics is 10,000 times as fast.

The cracking feedstock contains more or less of metals (mainly vanadium, nickel, and iron), sulfur and nitrogen, in addition to oxygen. Organometallic links are broken and their metals accumulate on the catalyst, leading to its deactivation, accelerated formation of coke, and higher amount of gaseous hydrocarbons.

Cracking of aromatic feedstock is characterized by an increased efficiency of the formation of aromatic hydrocarbons with a considerable admixture of olefins, in addition to the higher amount of coke. A naphthene feedstock produces a top quality gasoline as the result of isomerization and aromatization reactions.

A fluidized-bed catalytic cracking plant is composed essentially of a vertical-tube reactor, raising the catalyst and raw material (the basic process zone) and a regenerator with pipes carrying a spent and regenerated aluminum-silicon oxide system. At the bottom of the vertical tube, the strongly pulverized catalyst is mixed with the – nearly entirely evaporated – heavy hydrocarbon feed; cracking takes place as the feed flows upwards at the rate of (4...12) m/sec at a pressure in the range (0.8...1;5) bar, at a temperature typically in the range (480...530)°C. The cracking of heavy petroleum feedstocks is accompanied by the formation of coke: it accumulates on the catalyst, blocking its active sites. In such conditions, it is gravitationally carried into a regenerator to remove the coke by burning, typically at temperatures in the range (635...650)°C.

The naphtha cut from the FBCC plant is the main source of sulfur being carried into the final gasoline products during the blending process. In the global refinery industry, the level of sulfur in nahptha obtained by FBCC is reduced by the following methods:

- pre-treatment of the FBCC feedstock using a hydrogen-catalyst method for the removal of the entrained sulfur;
- increasing the conversion of organic sulfur compounds into hydrogen sulfide during FBCC;
- processing the FBCC product by distillation with absorption.

The highest percentage of sulfur is concentrated in the highest-boiling gasoline fraction from FBCC. Therefore, lowering the final boiling range of that fraction is the obvious method to reduce its sulfur content. The available techniques include the following:

- dropping part of naphtha into light diesel fuel; on the other hand, rejecting part of naphtha leads to higher quantities of light oil being collected but reduces its boiling range, changes the heat load of the light oil (part of which is recirculated within the main rectifying column) and degrades part of the naphtha to the medium distillate range;
- collecting separately the heavier naphtha cut as part of the overall distillation of FBCC products; however, this changes product proportions, operation of the major fractionating column, and operation of the gas absorption system.

Hydrogen-based Processes

Hydrogen-based processes are thermo-catalytic processes, carried out at free-hydrogen pressure conditions. There are three different variants of hydrogen-based processes, depending on the degree of conversion; only those relating to the production of fuel components will be discussed later in this chapter.

Hydrodesulfurization of The Cracking Feedstock

A hydrocracking plant usually comprises the following units: hydrocracking of vacuum distillates, hydrogen generation, and hydrogen recovery from post-production gas. The process is intended to handle vacuum distillates from the pipe-tower distillation system with a boiling range (330...575)°C and provide desulfurized products having lower molecular weights and lower boiling ranges.

The process of technology in the plant is divided into the following steps: hydrodesulfurization, hydrocracking, and fractionation of hydrocracking products. Hydrodesulfurization and hydrocracking take place in the presence of catalysts at elevated temperatures (340...450)°C, at a hydrogen pressure of about 15 MPa. The hydrodesulfurization reaction is accompanied by the removal of other contaminants from the feedstock (including nitrogen, chlorine, oxygen, metals), hydrogenation of olefins, and part of aromatic compounds. The main reason why the metals are removed is to protect the catalyst from irreversible deactivation taking place as metal compounds accumulate on the catalyst's surface.

The mechanisms of hydrocracking include two basic conversions: cracking of hydrocarbons, and hydrogenation of the products of catalytic cracking, typically in the presence of an aluminosilicate-based nickel-tungsten catalyst. The post-reaction mixture provides the following fractions: liquid gas, light gasoline, middle gasoline, aviation fuel, light diesel fuel, heavy diesel fuel, and a desulfurized but non-cracked vacuum oil fraction which is a feedstock to the fluidized-bed catalytic cracking (FBCC) plant.

Hydrocracking has the essential advantage of providing top quality products which, unlike similar products obtained by catalytic cracking, have a better stability because they contain no olefins or dienes. Moreover, the content of sulfur and nitrogen in the gasoline and diesel fuel products obtained is low enough to enable them to be used directly in obtaining environmentally-friendly blends of final products.

Destructive Hydrogenation

Destructive hydrogenation is a process in which a solid and a liquid feedstock is cracked under a hydrogen pressure in the range (29419.8...68646.2) kPa at a temperature in the range (420...500)°C in the presence of catalysts (Fe, W, Mo, Ni). The process is intended to produce gasoline products, but sometimes also diesel fuels from coal, bituminous shale, tar and soft asphalt.

Catalytic Processing of Gases and Light Gasoline Fractions

The processing of light fractions and gases is intended to provide saturated components of fuels or products for petrochemical syntheses. The process includes the following reactions, running in the presence of suitable catalysts:

- polymerization of gaseous alkenes;
- alkylation of gaseous and liquid isoparaffins with alkenes;
- alkylation of aromatic compounds with alkenes;
- dehydrogenation of butane and pentane fractions;
- isomerization of butane and light hydrocarbons from gasoline fractions.

Although rather energy-consuming, the catalytic processing of these gases enables elimination of distillation losses, that is, flare combustion of gases while increasing the obtained amounts of gasoline.

Efficient Processing of Soft Asphalt

Soft-asphalt utilization technology includes thermal methods (mainly coking and *visbreaking*), extraction, hydrogen-based methods (such as hydrodesulfurization and hydrocracking), and gasification.

Among these methods, hydrogen-based processes and gasification are believed to have the highest potential and be most environmentally-friendly, although they require the highest investment costs.

Lurgi offers a Multi Purpose Gasification (MPG) technology for gasification of hydrocarbon feedstock. Its main advantages include the possibility to handle low-quality/high-viscosity heavy fractions, also with a content of sludge, mud, and waste coke, and the possibility to handle raw materials with a high sulfur content. The oxygen-steam feedstock gasification unit comprises a burner and a reactor, gas cooling section, and a system for the removal of ash, metals, soot, and liquid waste.

Gasification is an autothermal process, controlled by the oxygen-to-steam ratio, running according to the following reaction:

$$2CH_n + O_2 \rightarrow 2CO + nH_2$$

$$CH_n + H_2O \rightarrow CO + (n/2 + 1)\, H_2$$

The ratio of CO and H_2 generated in syngas depends on the composition of raw materials, oxygen-to-steam ratio, and parameters of gasification. Non-catalytic semi-combustion of hydrocarbons in the MPG technology takes place in an empty reactor lined with a refractory material, selected for a load resulting from the ash content in the feedstock. The material is fed into the reactor through the burner at the top of the reactor. The burner accepts liquid feeds with the highest viscosities as well as emulsions and sludge with particles the size of several millimeters. The feed and an oxidizer are heated and mixed with steam as a moderator before the burner. The burner and the reactor are "fine-tuned", or adapted, to each other by dynamic simulation to entirely mix the reactants in as small a volume as possible and, in this way gasification of the raw materials is complete. The hot crude gas from the reactor is quenched with water originating from the ash and soot removal unit. The water is injected in a radial arrangement into the quenching-ring zone, where it is quenched, or cooled down rapidly, into the form of glassy beads the size of (1...2) mm. The beads accumulate at the bottom of the separator and are discharged through a hopper. The glassy slag carries a majority of heavy metal content and water-insoluble components. Further cooling takes place in a medium-pressure steam boiler, generating steam in the range (1.5...3.0) MPa. Final cooling takes place in a water cooler, then the gas is sent to the acid gas removal unit.

The crude syngas portion which is intended for use in hydrogen generator, passing by the steam boiler, is sent straight into the CO catalytic conversion unit, working according to the following reaction:

$$CO + H_2O \rightarrow CO_2 + H_2$$

Carbon dioxide is removed by means of cooled methanol.

The syngas, generated during gasification of hydrocarbons, contains a certain amount of free carbon (soot), typically about 0.8 % (m/m) on the feedstock basis. The soot particles are removed from the gas together with ash, mainly in the Venturi gas srubber, located after the quenching section. Sludge with a content of soot is collected together with condensates from the steam boiler and from the cooler-gas scrubber, and is sent to the ash (and metals) removal section. The sludge, with soot from gasification, is depressured batchwise to an atmospheric pressure in the sludge tank before being filtered.

The German refinery PCK Schwedt and Toyo Engineering Corporation from Japan developed one of the most advanced processes of technology for processing the residue from vacuum distillation of petroleum: HSC-DESUS (*HSC-High Conversion Soaker Cracking*). Compared with conventional visbreaking, it is characterized by a very high conversion and the residue has a stable quality. A variety of intermediate products can be used as a feedstock with a high content of sulfur and heavy metals (including heavy oil and bitumens from oil sands as well as residues from the production of lubricating oil). Post-cracking distillates from that technology are usually light or heavy gas oils having a lower content of unsaturated compounds than that in distillates from coking processes.

In the HSC technology, the feedstock is heated to a temperature in the range 440...460°C, depending on the desired conversion in the soaking drum. Cracking in the furnace is minimized by using high flow rates. A reactant stream from the furnace is made to flow into the soaking drum in which the residence time is long enough to provide desired conversions. The soaking drum operates at an atmospheric pressure, and its bottom section is filled with stripping steam. In the soaking drum, the raw material flows downward, through perforated plates. Steam along with cracking gas and distillate vapor flow through the perforated plates upward; their flow is countercurrent, compared with that of the raw material. The temperature in the soaking drum is highest at the top and becomes lower in its lower sections due to the adiabatic cracking reaction and stripping of the cracked substrate. The liquid from the bottom is pumped out and quenched in the heat exchanger to a temperature of less than 350°C. Vapor from the soaking drum flows into the rectifying column in which desirable intermediate fractions are formed, including heavy vacuum oil.

The soaking drum contains a stable homogeneous dispersion of asphaltenes in the residue, even at much higher conversions than in conventional visbreaking.

Distillates from conventional visbreaking, which is regarded as a first step in the processing of soft asphalt, and from the second step (the HSC plant) are then subjected to hydrogen treatment in the DESUS plant. The feed, after a multi-step heat exchange and after being mixed with hydrogen, is made to flow through the furnace and into a fixed-bed reactor. This is the hydrofining reactor, where sulfur, nitrogen, and oxygen are removed from the liquid reactants and hydrocracking takes place, with conversions of around 30 % (m/m). The post-reaction mixture flows through heat exchangers first into the hot separator, then the vapor and gas flow through a heat exchanger and two-step coolers into a cold separator for the separation of the liquid/gas phases. Liquid products are sent downstream to rectification, except that light fractions are subjected to stabilization.

MANUFACTURING OF SYNTHETIC LIQUID FUELS

The process to make synthetic fuels from syngas is known as the Fischer-Tropsch synthesis and was first used commercially in the 1940's. The process to make engine fuels from a gas which contains a mixture of carbon monoxide and hydrogen, in the presence of cobalt and iron catalysts runs according to the following reaction:

the cobalt catalyst

$$(2n + 1) H_2 + n\ CO \rightarrow C_n H_{2n+2} + n H_2O$$
$$2n\ H_2 + n\ CO \rightarrow C_n H_{2n} + n\ H_2O$$

the iron catalyst

$$(n+1)\ H_2 + 2n\ CO \rightarrow C_n H_{2n+2} + n\ CO_2$$
$$n\ H_2 + 2n\ CO \rightarrow C_n H_{2n} + n\ CO_2$$

The process comprises:

- syngas production by oxygen-steam gasification of coal, or by the catalytic semi-combustion or catalytic reforming (or both processes) of natural gas in the presence of steam;
- removal of sulfur and carbon dioxide;
- catalytic synthesis of carbon monoxide and hydrogen to form hydrocarbons;
- distillation and treatment of the resulting intermediate products to obtain: liquefied gas, gasoline, jet fuels, diesel fuel, and paraffins.

The first American plant using the Fischer-Tropsch technology was built, in 1951, in Brownsville, Texas, by Carthage Hydrocol Company. It was based on a desulfurized natural gas. The plant was composed of two reactors (diameter 5.1 m, height 24 m), packed with 200 Mg of a fluid-bed iron catalyst each. A simplified diagram of the process is shown in Fig. 2. The oxygen generating plant supplied 1800 Mg O_2 per day into the generator of catalytic semi-combustion of methane. The whole process was carried out at 3 MPa. Carbon dioxide was removed from syngas using a water jet.

Water, as a coolant, was made to flow through coolers located in the catalyst bed. Heat, generated during the reaction, was used for making steam. Gas and synthesis products were collected at the top of the reactor, carrying along the fine dust of the catalyst after removing it downstream by means of cyclones. The condensing portion of hydrocarbons and oxygen compounds was chilled and washed out with water. Lighter hydrocarbons were removed using an absorption-desorption system. C_3 and C_4 olefins were polymerized catalytically, obtaining gasoline which was then refined.

The resulting liquid products were composed of 25 % oxygen compounds and 75 % hydrocarbons. Gasoline after final treatment had a high olefin content and its octane number was 85.

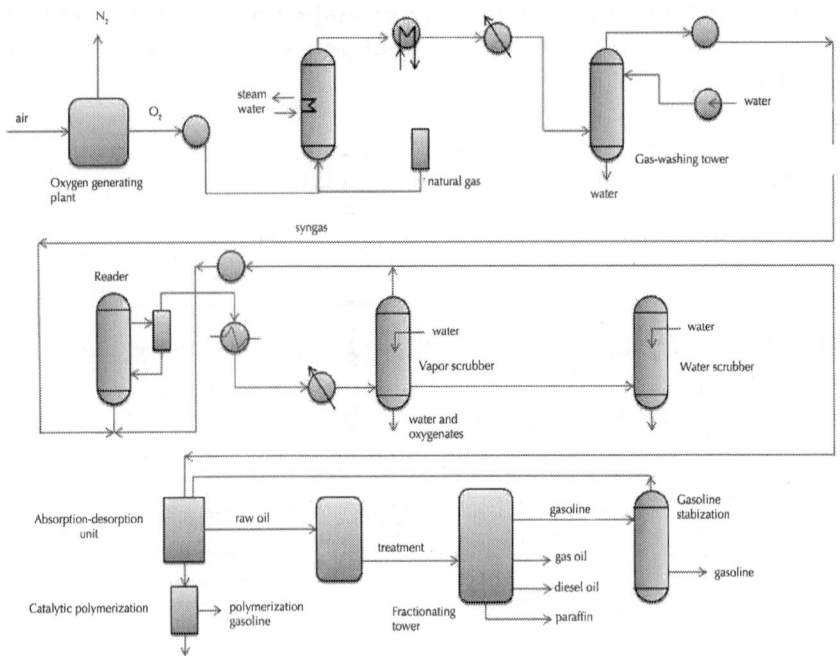

Figure 2: Simplified diagram of the Hydrocol plant.

In 1955, in Sasolburg, South Africa, a coal-based plant using the Fischer-Tropsch synthesis was started. The raw material, named SASOL, is a poorly-sintering type of coal, containing about 25 % ash and 10 % water, its heat of combustion is 23,000 kJ/kg. Fig. 3 shows a general diagram of the process to obtain synthetic fuels in the Sasolburg plant.

Coal is crushed into finer pieces and then divided into three categories. The smallest pieces are used in the power station in four boilers, each of a capacity of 160 tons of steam per hour. After purification, the gas at a pressure 2.5 MPa is separated into two streams: one stream is sent to the synthesis unit which is equipped with revamped fixed-bed cylindrical reactors (each has a heat exchanger, cooler and blower enabling the gas to be recirculated). The other gas stream is made to flow to fluidized-bed reactors which are supplied with the gas obtained from conversion of the C_1 and C_2 hydrocarbons being made at the same plant. Each reactor in the synthesis plant is provided with approx. 20,000 m³ of syngas per hour. The gas conversion is in the range (50...60) %.

A majority of the products obtained in the synthetic plant are high-boiling hydrocarbons. The fluidized-bed reactors produce mainly gasoline.

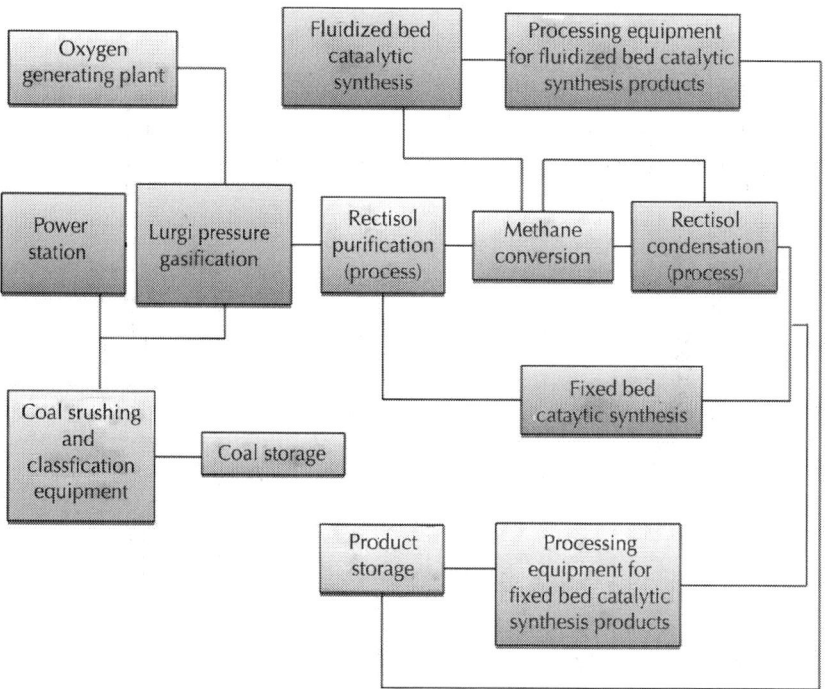

Figure 3: A block diagram of the synthetic fuel plant in Sasolburg (source: own elaboration on the basis of company materials).

Methanol as a Starting Material for Making Liquid Fuels and Petrochemical Products

At present, the synthesis of methanol is carried out globally starting from a mixture of carbon monoxide and dioxide and hydrogen in the presence of catalysts which typically contain Cu-Zn-Al or Cu-Zn-Cr at a pressure in the range (5...10) MPa at a temperature in the range 240...275°C, according to the following reaction:

$$CO + 2H_2 \rightarrow CH_3OH \qquad \Delta H^0 = -92 \text{ kJ/mol}$$

and, in part:

$$CO_2 + 3H_2 \rightarrow CH_3OH + H_2O \qquad\qquad \Delta H^0 = -50 \text{ kJ/mol}$$

The process selectivity is enormously high, therefore, only a maximum of 0.5 % (m/m) of side products (in addition to water) is made.

A diagram of the complete methanol plant, based on natural gas or biogas from a municipal waste dump or anaerobic biomass fermentation tanks, is shown in Fig. 4. The methane or biogas is heated in the central furnace to approx. 420°C and made to flow into a desulfurization reactor which is packed with zinc oxide beads ($ZnO+H_2S \rightarrow ZnS+H_2O$). From the reactor, the gas flows into the saturator (scrubber) to be saturated with steam. The saturator is supplied with a mixture of hot water from the distillation section and from the process condensate tank. Methane is heated in the respective apparatuses to about 800°C after being saturated with more steam (live steam) to form the desired mixture, and then directed into the reforming unit I° for the endothermal reaction $CH_4+H_2O \rightarrow CO+3H_2$ to take place on the Ni/ -Al_2O_3 catalyst. If a biogas with 60 % CH_4 and 40 % CO_2 is used for the synthesis of methanol, then the reaction $CH_4+CO_2 \rightarrow 2CO+2H_2$ takes place additionally. Approximately 10 % (v/v) of methane is not converted, therefore, the reaction mixture is sent into the reactor II° for a strongly exothermal semi-combustion of CH_4 to take place with participation of a strictly measured amount of oxygen on a nickel catalyst, according to the reaction:

$$2CH_4 + O_2 \rightarrow 2CO + 4H_2$$

A syngas at nearly 900°C is then made to flow through a boiler which generates compressed steam for the methanol plant, and is combined with hydrogen, separated from fuel gas (post-production gas) by Pressure Swing Adsorption (PSA).

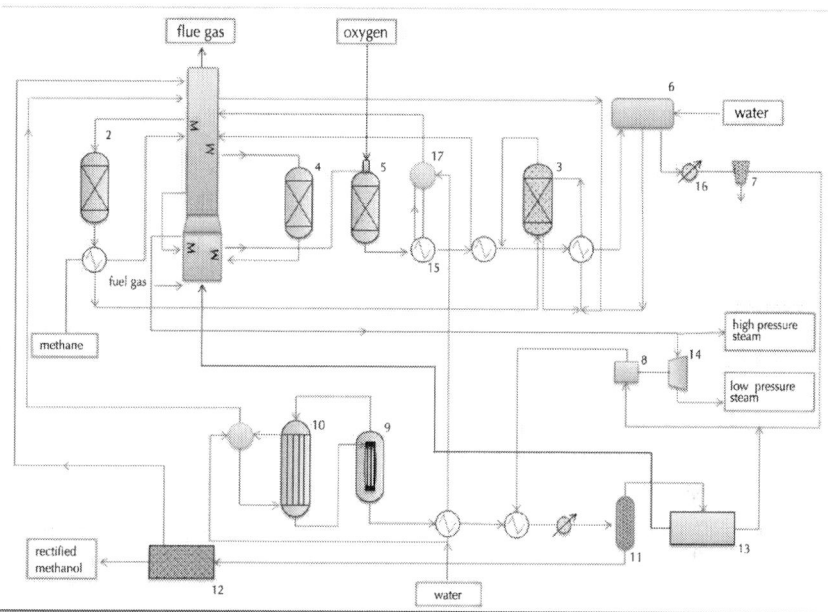

Figure 4: A diagram of the methanol plant based on natural gas or bio-gas at Lurgi Öl-Gas (source: own elaboration on the basis of company materials) 1 – central gas-fired heater, 2 – methane desulfurization reactor with a ZnO catalyst, 3 – saturator, 4 – stage 1 reactor for methane steam reforming over a Ni catalyst, 5 – stage 2 reactor for methane reforming with the use of oxygen, 6 – water evaporator, 7 – water separator, 8 – turbocompressor, 9 – stage 1 reactor for methanol synthesis, 10 – stage 2 reactor for methanol synthesis, 11 – methanol recovery from unreacted syngas,12 – 3-column distillation of raw methanol, 13 – hydrogen recovery from waste gases, PSA method,14 – steam turbine which drives the turbocompressors of syngas and recycle gas, 15 – heat exchanger,16 – cooler, 17 – water-steam separator.

Then, it is sent into the turbo-compressor to obtain a pressure in the range (5...10) MPa and is combined with the unreacted reactant being recirculated from the methanol separator, and sent to two serially-connected methanol synthesis reactors. After the second synthesis reactor, the post-reaction mixture flows through the heat exchanger and water-cooled methanol condenser into the product separator. From the bottom part of the product separator, a crude methanol flows through pressure reduction valves and into three serially-connected rectifying columns. After the last rectifying column in the series, the purity of the methanol product is a minimum of 99.99 %. A portion of the unconverted post-process gas is recirculated for another synthesis of CH_3OH, while another portion is sent through the PSA system (for recovery of hydrogen) into the fuel gas network and then into the central furnace. Bleeding a portion of the unconverted post-process gas continuously enables the inert content in circulation (N_2+CH_4) to be kept at a stable level of approximately 15 % (v/v).The first step, in which a crude methanol is neutralized using an aqueous soda lye solution, is followed by oxidation of aldehydes and iron cabonyls using an aqueous potassium permanganate solution. This results in the formation of a sludge which comprises manganese dioxide and other solids and is removed by filtration. The crude methanol is refined by distillation in two or three rectifying columns. For a two-column system, two process variants are possible: either both columns operate at an atmospheric pressure, or one operates at an atmospheric pressure while the other does at an elevated pressure.

The components of crude methanol are divided into three essential groups: light components, ethanol, and higher alcohols. Low-boiling compounds, having boiling points lower than or close to that of methanol, are removed along with the dissolved gases from the synthesis plant (H_2, CO, CO_2, CH_4 etc.) in the first column. It is also referred to as the extraction column, with water being the extraction agent, which differentiates the liquid phase activity coefficients of the major components. The scope of changes in the activities of the respective components is determined by the quantity of the solvent required.

The crude feed is pre-treated with chemicals (NaOH, $KMnO_4$), then heated to a boiling range and made to flow into the middle section of the extraction column. Extraction water, which is recirculated from the bottom section of the second rectifying column, is fed to the top of the extraction column. When extracting, the water is cooled down in the heat exchanger by the flowing crude methanol feed. Vapors from the extraction column, flowing upwards, carry along all the volatiles and a small amount of methanol vapor with water. Most of the methanol, water, and of less volatile components are condensed in the condenser and recirculated as the reflux through the separator and into the extraction column, the other ones escape to the fuel gas network or to the flare.

After extraction distillation, the crude methanol from the column bottom contains roughly 50% (m/m) CH_3OH. The volume of the extractant to be added depends on the amount of contaminants as actually found in the crude feed and as shown in product specifications.

The second rectifying column is called the refining column: its main function is to separate water and high-boiling compounds from methanol being the principal product. Higher alcohols have a maximum concentration at the bottom of the column, below the feed point. Therefore, that part of the column comprises alternative points for carrying out higher alcohols from the system. The higher alcohols, if being carried out with the vapor stream, must be cooled down, condensed and sent into the separator. In addition to the higher alcohols and high-boiling components, the stream has a content of methanol, water, and small amounts of ethanol. If the stream is carried out from the lower trays of the stripping section or as a "liquid blow", it may have a significant water content and is likely to become separated into two phases if allowed to stand. The light phase (or the organic phase) is a quite valuable fuel and may be used as a fuel for a furnace (methane reforming reactor, or steam boiler). The aqueous phase may be drained and disposed of as a chemical liquid waste.

In a two-column process system, the pure methanol is collected at the top of the refining column. Typically, it is collected between trays number 4 and 6, counting from the top down. That section has 4 to 6 trays at the top (pasteurization section) and is designed to concentrate low-boiling components, which penetrate through the bottom of the extraction column. A small volume of the total condensate, which is obtained at the top, is intentionally "blown out" or recirculated into the first column for such trace amounts of minor contaminants to be finally taken to the top of the system and out.

With the second column operating as a pressure column, part of the vapor being collected at the top may be used to provide heat to the first distillation column. However, at higher pressures, the difference between relative volatilities for the respective components is lower, therefore, for same product purities and same heat loads in the evaporator, the pressure column needs more trays.

In a three-column methanol rectification system, the third column is designed to separate methanol from ethanol. One of the methods to remove ethanol is by "blowing out" a small stream above the feed tray in the refining column, or by collecting an ethanol-methanol mixture at the top of the refining column, followed by separation of the two components in the third column. The feed to that column contains nearly all of the ethanol present in the feedstock (crude methanol) plus the methanol product. Water and trace amounts of higher alcohols are removed from the column together with ethanol while pure methanol is collected overhead. The pasteurization section is located above the product collection point.

PRODUCTION OF GASOLINE

Gasoline is a hydrocarbon mixture of roughly 100 of various compounds, obtained by the straight-run and destructive processing of petroleum. The fractions, found in gasoline, include the paraffin fraction (40...65)% (V/V), naphthene fraction (20...35)% (V/V) and aromatic fraction: (8...20)% (V/V).

Taking into account the desirable method of combustion, gasoline should preferably contain large amounts of aromatic hydrocarbons, obtained mainly by reforming and partly by pyrolysis. Very important components – especially in aviation fuels – are mixtures of hydrocarbons obtained in isomerization and alkylation processes (isoparaffins). Highly resistant to knocking combustion, they have the added advantage of a sufficiently high calorific value.

All the components that are required for gasoline blending are obtained in the respective types of refining processes and are combined in the blending unit in accordance with process requirements for the specific gasoline types. Major components of gasoline are listed in Table 1. The respective components are blended so as to obtain a final product (commercial gasoline) which complies with the requirements of applicable standards.

Most of those qualitative parameters of fuels are only approximate because their fractional and chemical composition is very complex. Numerical values for the respective qualitative parameters are selected in a manner which, in as far as possible, enables compliance with performance requirements, resulting from the design and mode of operation of spontaneous-ignition engines fuelled by such blends.

Table 1: Characteristics of gasoline components

Component	Boiling point/ range [°C]	RON	Function or use
Base cut from distillation	45 ÷ 195	40 ÷ 54	Essential component of gasoline
Butane	0	95	Facilitates cold engine start, component of automotive gasoline
Pentane-hexane fraction	27 ÷ 65	45 ÷ 93	Enables continuous combustion during startup, when present in automotive and aviation gasoline types
Light distillate	65 ÷ 90	-	same as above
Reforming product: - complete	-	-	Major component of automotive and aviation gasoline types, resistant to detonation combustion
- de-aromatized	45 ÷ 200	92÷ 101	
- de-xylenated	-	-	
- refined reformate	-	-	
Hydrocracking fraction	40 ÷ 200	72 ÷ 85	Component of automotive and aviation gasoline types with good anti-detonation properties and low sulfur content
Catalytic cracking fractions	40 - 200	91 – 93	Widely used as a component of high-octane gasoline
Hydrogenated pyrolysis gasoline	60 ÷ 200	96 ÷ 99	Used in smaller amounts, especially for automotive gasoline
Polymerization gasoline	60 ÷ 200	94 ÷ 96	Less important, high-octane component of gasoline
Isooctane (2,2,4-trimethylpentane)	111	100	High-octane component of aviation fuels
Various alkylates	40 ÷ 150	93 ÷ 96	Widely used in blends of aviation gasoline (usually), and automotive gasoline (less frequently)
Various isomerisates	40 ÷ 70	82 ÷ 85	Desirable as components of gasoline types for high-duty applications
Additives	-	-	Component which improves lubrication and resistance to detonation combustion, depending on its essential composition

The quality or properties of fuels affect a number of processes, connected with their use in a wide sense. The impact of the most significant qualitative parameters of fuels for spark-ignition engines on the various performance processes has been established in a number of tests. The criteria are shown inTable 2.

Table 2: Qualitative criteria of gasoline in their performance processes

Requirements relating to gasoline				
Properties of fuels	Storage, distribution, fuelling	Formation of fuel-air mixture	Optimum combustion	Environmental impact
	Density	Chemical composition	Chemical composition	Formation of toxic components of emissions
	Chemical stability	Fractional composition	Fractional composition	
	Corrosive effect	Vapor pressure	Calorific value	
	Level of contamination	Heat of evaporation	Resistance to detonation combustion	
	Low-temperature properties	Washing properties		Biodegradability
	Electrostatic properties	Density		
	Fire safety	Viscosity		

OBTAINING OF DIESEL FUELS

Diesel fuels are made by blending fractions having boiling ranges from 190°C to 350°C, obtained from petroleum processing in the following technologies:

- an oil fraction from atmospheric distillation (base fraction) of which the properties depend on the chemical nature of petroleum;
- an oil component obtained by thermal cracking; it has a low cetane number, low stability, and a considerable content of unsaturated hydrocarbons;
- an oil component obtained by catalytic cracking, which has a rather low cetane number (CN=40...60) because of a content of aromatic hydrocarbons;
- an oil component obtained by hydrocracking, used for reducing the corrosive effect of diesel fuels by decomposing sulfur compounds in the process; the amount of the hydrocracking product component which is present in the finished diesel fuel determines the oil category in respect of its content of sulfur links;

- an oil component obtained by dewaxing; it deteriorates spontaneous-ignition properties of oils while much improving their low-temperature properties;
- a light oil fraction obtained by vacuum distillation (depending on the grade of petroleum being processed) which increases the amount of oil product thus reducing its volatility and increasing its viscosity and density.
- Depending on the intended use, diesel fuels are blended essentially in two groups:
- diesel fuels for high-speed engines;
- diesel fuels for medium-and low-speed engines; another type of blends comprises so-called heating oils which are intended for use in steam boilers for marine or land applications, industrial furnaces (in rolling mills, glass works etc.), firing up coal dust-fired steam boilers, and for technological reasons.
- The following improvers can be added to diesel fuels, depending on their intended application:
- pro-detonators – to increase the cetane number of diesel fuels;
- corrosion inhibitors – to reduce the corrosive effect of diesel fuels and their combustion products;
- oxidation inhibitors – to improve diesel fuels in terms of stability, enable longer storage;
- depressants – to lower the freezing point of diesel fuels;
- additives which reduce the smoke level in exhaust gases by improving the combustion process.

Those fuels for spontaneous-ignition engines (diesel fuels) which have the desirable composition are expected to show the following characteristics:

- ensure the correct functioning of the fuel system, especially the injection assembly;
- ensure a correct and energy-efficient combustion;
- ensure reduction of toxic components and solid emissions;
- guarantee chemical stability in the storage process.

The qualitative criteria for diesel fuels which are important for the whole area of their application are shown in Table 3.

Table 3: Qualitative criteria of diesel fuels in their performance processes

Requirements relating to diesel fuels				
Properties of fuels	Storage and distribution	Functioning of fuel distribution system	Atomization, evaporation, and combustion	Environmental impact
	Density	Density	Viscosity	Formation of toxic components of emissions
	Chemical stability	Viscosity	Surface tension	
	Low-temperature properties	Low-temperature properties	Spontaneous ignition properties	
	Corrosive effect	Lubrication	Spontaneous ignition properties	
	Resistance to microbial contamination	Level of solid contaminants and water		Biodegradability
	Electrostatic properties		Calorific value	
	Foaming			
	Fire safety.		Washing properties	

Regardless of standard fuels, for which applicable standards and approved testing methodologies exist, a number of alternative fuels are known, including biofuels, which can be applied as propulsion materials. Such fuels are used in typical spontaneous-combustion engines, of which the designs are adapted to the properties of conventional (standard) fuels. Therefore, taking into consideration engine requirements, the scope of the respective evaluation criteria ought to correspond to those which apply to conventional fuels. Existing alternative fuels for spontaneous-combustion engines for various applications are listed in Table 4.

Table 4: Alternative fuels for use in engines

Alternative fuels for use in engines			
Form	For spark-ignition engines	For spontaneous-ignition engines	For stationary engines
Liquid	methanol		
	ethanol		
	butanol		
	other alcohols (*tert*-butyl TBA, *sec*-butyl SBA, isopropyl IPA, neopentyl-NPA);	fatty acid esters (FAME, FAEE) from transesterification of rapeseed, soy, sunflower oils	
	Ethers (ethyl-*tert*-amyl TAEE, ethyl-*tert*-butyl ETBE, methyl-*tert*-amyl TAME, methyl-*tert*-butyl MTBE, diisopropyl DIPE);		tall fuels (TPO-tall pitch oils) obtained by esterification with ethyl / methyl alcohols of tall oils obtained from gums/resins of coniferous trees (side products in sulfate cellulose production processes and low-temperature dry distillation of wood);
	hydrogen-based synthetic fuels (including BG, FT, HTU processes)		
	liquefied petroleum gas (LPG)		
	dimethylofuran (DMF)	fuel-water emulsions (aquasols)	
		pure vegetable oils	
	liquefied natural gas (LNG)		
Gaseous	compressed natural gas (CNG)		
		biomethane from biogas	biogas
		dimethyl ether (DME) and (contemplated) diethyl ether (DEE);	
		gaseous fuels from CtG processes	
	hydrogen		

In view of the above data, for a rational assessment of the quality of fuels and their usefulness, especially after storage processes, in engine operation it is necessary to chose applicable assessment criteria and methodology, enabling a relatively fast analysis of changes in the parameter values. The choice of such criteria ought to result from the sensitivity of the respective criteria to fuels' oxidation and contamination, potentially causing the accepted and recognized limiting values to be exceeded both in respect of their measure and weight.

Experience in using engine fuels indicates that their quality may change, mainly in storage and distribution processes. This causes the necessity to establish the scope and frequency of the quality surveillance of fuels. Some of the generally adopted types, scopes, and frequencies of control of the quality of fuels in their distribution chain are shown in Table 5.

Table 5: Fuel types and scopes of analysis in the distribution chain

Fuel type and scope of control	Fuel distribution step
Full – comprising assessment of the values of all quality parameters of fuels, as described in the standard ON-EN 228 for gasoline and PN-EN 590 for diesel fuels	• Refinery – before delivering a fuel lot for distribution; • Storage facilities – after acceptance of fuel for storage and on its release, or periodically, every 6 months of storage
Control – comprising assessment of selected parameters, usually appearance, density, fractional composition, and vapor pressure – for gasoline, or flash point and cold filter plugging point – for diesel fuels.	• Storage facilities – periodically, during storage; • Fuel station – random analysis, for instance every 3 to 4 deliveries, after acceptance of fuel for storage.
Short – comprising determination of density, fractional composition, content of water and contaminants – for diesel fuels, or appearance, color, density – for gasoline.	• Fuel station – before acceptance of fuel for storage; • Storage facilities – before unloading tankers.

CONCLUSIONS

The growing demand on liquid fuels necessitates maximization of production output, especially those fuel components which originate from destructive processing. Even though in straight-run processing of petroleum, processes may be conducted which are intended to expand the limits of fractions of base gasoline, kerosine and diesel fuels fractions, yet the main focus is on secondary processing, providing increased numbers and amounts of components of diesel fuels, also by means of thermal and thermocatalytic processes, in the presence or absence of hydrogen.

Owing to the growing number of spontaneous-ignition engines in Europe, the supply of diesel fuels is insufficient while that of gasoline is excessive. As a result, technological processes are carried out which provide the maximum yield of propellant cuts and the residues are processed to provide components which are useful for diesel fuel blending processes.

Experiments were made in which fractions resulting from depolymerization of plastics (KTSF fraction) were "sunk" in petroleum or components which result from re-refining of spent lubricating oils were utilized.

In the production of gasoline, if correct process conditions are maintained, components obtained in isomerization processes, catalytic reforming, full hydrocracking, alkylation (using isobutane) and fluidized-bed cracking (FBCC) are not expected to affect the stability of gasoline during long-term storage.

As regards the motors spirits manufacturing and blending technologies, the following fractions, which originate from the processes of technology discussed above, may very much reduce the duration of safe storage of such fuels:

- the gasoline fraction obtained by thermal cracking of the vacuum column residue, which was not hydrogenated;
- alternately, the non-hydrogenated fraction of pyrolysis gasoline;
- fractions from synthesis of gaseous hydrocarbons;
- ethanol as a biocomponent.

The contemporary diesel fuel blending techniques are typically based on the combining of components derived from the following major process unit:

- distillation in a tube-tower distillation system
- hydrocracking;
- fluidized-bed cracking (FBCC);
- hydrodesulfurization of soft asphalt;
- thermal processing of residue.

The stability of diesel fuels can be much affected by the following factors:

- components from thermal processes;
- biocomponents (FAAE);
- components from WtL processes (KTSF fraction) and re-refining products of the processing of spent lubricant oils.

Some refineries offer co-hydrogenation of petroleum fractions and vegetable oils. The solution carries a potential risk to the blend stability because the process mechanism, connected with the presence in such oils of heterogenic compounds, is not very well known.

Generally, the stability of fuel blends is the lower, the more unsaturated bonds such as those in alkenes (or olefins) they contain.

REFERENCES

1. K. Biernat, „Kierunki rozwoju układów zasilania i spalania oraz ródeł energii we współczesnych silnikach spalinowych", [*Trends in the development of fuelling and combustion systems and energy sources in contemporary internal combustion engines*], „Przemysł Chemiczny" December 2006;

2. K. Biernat „Prognoza rozwoju paliw" [*A forecast for the development of fuels*], „Studia Ecologiae et Bioethicae" No. 3/2005;

3. K. Biernat, A. Kulczycki „Prognoza rozwoju biopaliw w Polsce", Krajowa Konferencja „Rynek biopaliw w Polsce – szanse i zagro enia", [*A forecast for the development of biofuels in Poland,* National Conference *The biofuels market in Poland – opportunities and risks*], Warsaw, 26.04.2006;

4. K. Biernat, A. Kulczycki, M. Rogulska: „Development of Alternative Fuels in Poland", conference proceedings: „Expert Workshop on Biofuels Support in Baltic States, Nordic Countries and Poland", Tallin, Estonia 14.02.2007.

5. J.M. Ma kowski, The EU Directives on Fuels Quality: a look beyond year 2000", conference proceedings: Mat. II Mi dzynarodowej Konferencji pt: "Rozwój technologii paliw w wietle Dyrektyw Europejskich i Narodowych Uregulowa Normatywnych" [2nd International Conference: The development of fuel technology in the light of European Directives and national normative regulations] Warsaw 1999

6. J. Gronowicz „Ochrona rodowiska w transporcie l dowym" [Environmental protection in land transport], ITE Radom 2004.

7. J. M. Ma kowski, „ Paliwa samochodowe pocz tku XXI wieku" [Automotive fuels of early 21st century], Wyd. Centrum Badawczo-Szkoleniowe Diagnostyki Pojazdów Samochodowych", Katowice 2002

8. K. Biernat "Materiały P dne i Smary" [Propellants and lubricants], WAT, Warsaw 1983

9. K. Biernat, Z. Karasz "Paliwa płynne i ich u ytkowanie" [Liquid fuels and their use], SIMP, Warsaw 1997.

10. K. Baczewski, T. Kałdo ski, „Paliwa do silników o zapłonie samoczynnym" [Fuels for self-ignition engines], WKiŁ, Warsaw 2004.

11. K. Baczewski, T. Kałdo ski, „Paliwa do silników o zapłonie iskrowym" [Fuels for spark-ignition engines], WKiŁ, Warsaw 2005.

12. A. Podniało, „Paliwa, oleje i smary w ekologicznej eksploatacji" [Fuels, oils, and lubricants in their environmentally-friendly operation], WKiŁ, Warsaw 2002.

13. D. Singh, „Meeting diesel specifications at sustained production", Hydrocarbon processing No. 4. 2011.

14. W. Kotowski "Maksymalizacja produkcji paliwa dieslowskiego w rafinerii przerobu ropy" [Maximization of production of diesel fuels in a petroleum refinery]. Paliwa Płynne No. 4 and No. 5. 2012.

Chapter 6

Criteria for the Quality Assessment of Engine Fuels in Storage and Operating Conditions

Krzysztof Biernat[1,2]

[1]Department of Fuels, Biofuels and Lubricants, Automotive Industry Institute, Poland

[2]Institute for Ecology and Bioethics of CSWU, Poland

INTRODUCTION

Plenty of literature exists on research and findings relating to changes in the performance characteristics of conventional fuels concerning their oxidation (ageing) processes. These processes affect a majority of the functional properties of fuels and cause them in many cases to exceed the values stated in applicable standards. However, the proper fuel storage time, that is, duration of the effect of many different factors on the process kinetics of low-temperature oxidation of fuels, is yet not clearly defined. The process kinetics is affected by a number of factors (including random ones), namely the following:

- fractional composition of fuels;
- structural-group composition;
- volume;
- surface area of the fuel mirror;
- free volume over the fuel mirror;
- type of tank;
- type of tank roof;
- tank foundation;
- type of tank layer material in contact with fuel;
- multiplicity of warehouse operations;
- tank breathing system;
- weather conditions.

Regardless of the above mentioned statements, fuels, being compositions of many types of hydrocarbons and their oxygen derivatives, may differ from one another in respect of structural-group composition within the various product lots of the same fuel grade, therefore, they are subject to ageing at different rates.

In light of the above, the correct organization and functioning of a fuel quality control system during storage is one of the criteria that enable fuel storage.

The fuel quality surveillance system should specify in particular:

- conditions of acceptance of the respective deliveries of product lots;
- scope and methodology for the quality control of the accepted product lots;
- duration of safe storage of fuels, pre-determined from the results of assessment of selected parameters, which are significantly affected by the anticipated conditions of storage;
- the duration and scope of control analyses (full, short, basic) and their definitions;
- how to deal with products of which the properties are outside the ranges referred to in the appropriate standard specifications;
- a list of documents required, and templates;
- authorizations and responsibilities of the staff with regard to maintaining the quality of stored fuels.

As the requirements relating to fuels quality may change and progress in their manufacturing technology may be observed, assessment of the criteria of fuel quality during storage may be modified accordingly. Hence, it is necessary to review such assessment criteria and amend them as necessary, so as to enable the assessment of changes in the quality of fuels on a regular basis, since the results are meant to help take decisions on how to handle the fuel stored.

The choice of criteria for assessing the quality of fuels should, therefore, be dictated by the requirement to determine at certain time intervals only those parameters of which the values may change due to the low-temperature oxidation of fuels, which occurs during their storage. The right choice of such parameters will be possible after analyzing the changes observed in the properties of standard fuels during storage and, as a result of such analysis, determination of the proper hierarchy of assessment criteria and their weights and measures; these issues are going to be discussed later in this chapter.

Evaluation of the usefulness of fuel delivered for storage and forecasting the maximum duration of safe storage in specific storage capacities are an important element of the effective operation of the Quality Assessment System. Therefore, a full analysis of the received product is required, in addition to a certificate from the fuel supplier's laboratory. The certificate typically includes the results of fuel quality assessment made by the manufacturer at the time of manufacturing. The time that elapses between the production of a fuel lot and its intermediate storage and distribution processes may result in significant changes in the quality of fuel, affecting the process of its storage, all the more so as fuel handling habits may not always be appropriate.

If the full analysis results show limiting values, in particular those which may be subject to change due to the storage process (e.g.: vapor pressure, resin content, oxygen content, fractional composition for diesel fuels, flash-point, solids, etc.), the available options include the refusal to accept such fuel for storage, or its acceptance on the supplier's responsibility.

In order to ensure compliance with the relevant fuel storage procedure, in the case of fuels intended for long-term storage, a reserve in the range of properties that can be affected in the normal execution of the process should be provided. Hence, as in some countries, a proper fuel storage standard should be prepared and implemented.

The standard should take into account a "quality reserve" with regard to those parameters which are particularly sensitive to low-temperature oxidation.

CRITERIA FOR AN ASSESSMENT OF THE QUALITY OF GASOLINE IN STORAGE AND DISTRIBUTION CONDITIONS

In the classification of combustion engines, spark-ignition engines are in the group of engines with external preparation of the fuel mixture. These engines are forced-ignition engines and the fuel mixture is prepared outside the combustion chamber; this means that the necessary condition is to generate a mixture of fuel vapor and air. Evaporation is defined as the passage of liquid into the gaseous state. Evaporation occurs practically at any temperature in liquids having a free surface. Evaporation is of the free type (without any factors forcing the motion of particles, i.e., into a motionless medium) and of the forced type, i.e., into a moving medium.

The molecule of a liquid passes into the gaseous state after overcoming the forces of intermolecular cohesion – therefore, it does a certain work (E). The kinetic energy of the molecule must be greater than the work, hence:

$$\frac{m\,u^2}{2} \geq E$$

(1)

where:

m-mass of the molecule;

u – velocity component, perpendicular to the liquid surface.

The movement of molecules takes place in both directions, i.e., the molecules also pass from the gaseous into the liquid state. Depending on the number of particles passing into the vapor state (n) and into the liquid state (m), the following phenomena may occur:

n> m-evaporation;

n=m-saturated vapor state;

n <m-condensation

The condition in which surface evaporation takes place in addition to volumetric evaporation is important for the evaporation process. In this case, gas bubbles are produced throughout the liquid volume and are directed towards the liquid surface. This rapid evaporation is called "boiling" and, for a homogeneous liquid, is characterized by a constant temperature from the beginning of boiling till the whole liquid has evaporated. Gasoline, a mixture of many components, is characterized by a boiling range which depends on the concentration and boiling points of its individual components.

The rate of evaporation is affected by a number of different factors, including external ones, such as the following:

- *temperature*– increase in temperature causes a rise in the kinetic energy of molecules, and therefore an increase in temperature causes an increase in the intensity of evaporation, up to the boiling condition;

- *pressure*– the lower the pressure in the medium into which the liquid evaporates, the higher the intensity of the evaporation process taking place. The phenomenon is used in combustion engines, where the suction stroke in the cylinder generates partial vacuum which, in turn, facilitates fuel evaporation. Thus, reduction in pressure causes a decrease in the boiling range of the liquid; this is used in vacuum distillation.

- *concentration difference*– removal of the vapors, accumulated above the liquid surface increases the gradient of concentration and accelerates evaporation of the liquid;

- *diffusion coefficient* – specifies the amount of mass transferred in a defined unit of time per unit of area; thus, the higher the diffusion coefficient, the more intense evaporation of the liquid will occur;

- *surface area of the liquid.*

Vapors from above the liquid surface can be removed by diffusion. The phenomenon consists in that the solute (solved in any solvent) tends to be evenly distributed in the entire volume of the solvent. In this case, the volume of air (which is limited by the size of the combustion chamber or an open space in the case of free evaporation) is the solvent for the fuel vapor. Therefore, fuel vapors perform in the process of diffusion a series of thermally excited, small transitions. The process of diffusion is determined using Fick's first law:

$$\frac{dn}{dt} = -Ds\frac{dc}{dx}$$

(2)

where:

n – number of moles of molecules which diffuse in time "t"

s – distribution area surface

D – diffusion coefficient

$\frac{dc}{dx}$ -concentration gradient in the direction of diffusion

The rate of diffusion depends on temperature, pressure, and other factors.

From Fick's formula, after conversion, changes in the mass of the diffusing substance can be found from the following relationship:

$$\Delta M = -D(\frac{dc}{dx})ds\ dt$$

(3)

The diffusion process is characterized by the diffusion coefficient, which is equal to the number of moles of gas (fuel vapor) passing across 1 m² of the interface per second, when concentration – in the perpendicular direction to the interface – changes by one unit per 1 meter of the distance. Hence, the amount of mass moving during a unit of time across a unit of area depends on the value of the diffusion coefficient for the substance. Assuming that the diffusing particles have the shape of spheres – with the radius r – the relationship between the diffusion coefficient and viscosity is described by the formula:

$$D = k\frac{T}{6\pi\ r\eta}$$

(4)

where:

k-Boltzmann constant

T-temperature

η-dynamic viscosity

Taking into account the approximate relationship between the diffusion coefficient and viscosities on the one hand and the average velocity of vapor particles u_a and free path length τ on the other, we have:

$$D = \frac{1}{3} \tau \, u_a$$

(5)

and

$$\eta = \frac{1}{3} d \, u_a \tau$$

(6)

where:

d-density

it is possible to determine the estimated value of the diffusion coefficient D at a given temperature from a known viscosity and density, using the formula:

$$D = \frac{\eta}{d}$$

(7)

In addition to the factors mentioned above, fuel evaporation is affected by some internal factors, resulting from the chemical structure of fuels, among which volatility and heat of evaporation are the most important ones.

Heat of evaporation is defined as the amount of heat required to convert one unit of mass or one unit of volume of a liquid into the vapor state in specified temperature and pressure conditions. As the temperature changes into the critical point (CP), heat of evaporation gradually decreases, assuming the value of zero at that point. The heat of evaporation depends on the chemical structure (molecular mass) of the liquid and on the interaction forces (association) between the particles. Heavier fuels have higher heats of evaporation and tend to be evaporated with more difficulty, therefore, the use of conditions in which forced liquid evaporation takes place is required to obtain the appropriate composition of the fuel mixture.

In spark-ignition engines, intense evaporation of the fuel begins as the fuel leaves the nozzle. The fuel drops get disaggregated, forming tiny drops in the air jet stream. The phenomenon is called fuel atomization.

Fuel drops, which are suspended in the air and move around, evaporate gradually. As mentioned before, the rate of evaporation depends on pressure, temperature, flow rate of air, volatility, evaporation heat, diffusion coefficient, and the fuel drop surface. A portion of the fuel drops settle on the suction duct walls and gradually evaporate from there. In this case, intensity of evaporation largely depends on fuel drops' wettability, as compared with the suction duct material, expressed by the wetting angle. The larger is the wetting angle, the quicker the fuel's evaporation. However, at the same time, some of the fuel that settled on the walls in the form of liquid will be stripped mechanically by the steam of air and carried (in a non-evaporated form) into the combustion chamber. The phenomenon is applicable to the heavier components of fuels, also called the phlegm. The presence of the phlegm in the combustion chamber is an undesirable phenomenon, which affects the process of fuel combustion.

Roughly a half of the fuel evaporates in the suction duct, therefore, the composition of the fuel mixture in multi-cylinder engines is different in each cylinder. The differences in the blend composition can be up to a dozen or so per cent. Differences are seen not only in the fuel share but also in the chemical composition of the vapors and the non-evaporated stage in each cylinder. For obvious reasons, this affects the combustion process as well as the engine power and fuel consumption.

Part of the fuel gets into the combustion chamber in liquid form. Such fuel settles on the walls of the combustion chamber and piston head, and evaporates – mostly during the intake stroke and partly during the compression stroke. The least volatile components can evaporate as early as during combustion. If the fuel has a content of high-boiling components, they can decompose thermally instead of vaporizing, despite the high combustion process temperature; this leads to an immensely high contamination of exhaust gases, and to the formation of carbon deposits (solid decomposition products) on the chamber walls and piston head. For high rates of the suction/ compression/work processes, part of the fuel can also be removed from the chamber in liquid form, washing the lubricating oil from the walls, thus accelerating the wear-and-tear of the piston-crankshaft system. In that case, the fuel is removed from the combustion system along with the exhaust gas.

The elementary carburetor does not assure the correct supply of fuel to the engine when a fuel-rich mixture is required. Therefore, a series of devices were introduced to ensure a desirable composition over a wide range of operating conditions; they include starting devices, accelerating devices, and so-called "fuel savers". Unfortunately, these devices tend to complicate the carburetor's design, affecting its reliability and causing the need to adjust the carburetor as necessary.

Gasoline-fuelled engines with direct fuel injection by means of an injection pump do not have the above mentioned disadvantages in the carburetor system. This is because the mixture is evenly distributed into the respective cylinders and the adverse effect of the phlegm is also eliminated. However, technologically, direct injection is a much more complex system which requires extreme precision of workmanship and highly specialized repairs.

As a result, volatility is a very important operating parameter of fuels; it is estimated from their fractional compositions and vapor pressures.

Combustion is described as violent oxidation during which a high amount of heat is generated in addition to gaseous reaction products which create a flame. Combustion is one of oxidation reactions even though it does not require the presence of pure oxygen. Oxygen is the oxidizing agent in most combustion processes, other ones include HNO_3, H_2O_2, or F_2. The mechanism and the effect of combustion of hydrocarbons are identical regardless of the type of oxidizer. There are a wide range of compounds with a potential as fuels, however, most fuels are organic substances and their mixtures, containing mainly carbon, hydrogen and smaller quantities of nitrogen, sulfur and oxygen.

The complete combustion of a hydrocarbon fuel can be described by means of the following simplified reaction:

$$[nC, mH_2, kS] + pO_2 \rightarrow nCO_2 + mH_2O + kSO_3 + Q \tag{8}$$

Usually, the nitrogen content of fuels is not reactive. Incomplete oxidation may give CO and SO_2 as additional combustion products. Heat energy (Q) is generated as the bonds between atoms in fuel particles are broken and bonds in the reaction products are created. The higher the bonds in the combustion reaction products, compared with the bonds in the reactants, the higher is the energy effect of the reaction.

The temperature of the combustion process is influenced by dissociation of combustion product particles, which has a major role at temperatures above 2500°C.

Combustion is an exothermal reaction during which heat is generated. Activation energy (usually provided by heating) is needed to initiate combustion. The energy generated by the system is higher than the energy used for breaking bonds in the reactants. In engines having a carburetor, the energy needed for initiating combustion (also called "ignition energy") is provided by the ignition plug discharge and heating of the combustion chamber walls. In gasoline-fuelled engines, the ignition energy should be as low as possible, since such conditions lead to higher reaction heat effects.

After a flammable mixture is introduced into the combustion chamber but before it is ignited there elapses a period of time, called ignition delays. It is needed for the evaporation of gasoline particles getting into the combustion chamber in the liquid form and for their thorough mixing with air (practical delay), as well as for the initial, flameless oxidation of the gasoline vapors in the air so as to enable the formation of intermediate oxidation products which facilitate combustion of the mixture at a stable rate (chemical delay). The intermediate oxidation products include such compounds as: alcohols, aldehydes, ketones, acids and hydroxyacids. The initial oxidation process is called low-temperature combustion. In fuels with higher oxidation stabilities, fewer components participate in low-temperature combustion. According to Hess's law, the thermal effect of a chemical reaction depends on the initial and final states, rather than on transformation path. Although numerous intermediate combustion products are made in fuel combustion, its is safe to assume for the purposes of thermochemical balance that the initial state is: carbon and hydrogen, while the final combustion products are: H_2O, CO_2, SO_3. The general equation of the fuel combustion process can be formulated as follows:

$$C + O_2 = CO_2 + Q$$
$$2H_2 + O_2 + 2H_2O + Q$$
$$2S + 3O_2 = 2SO_3 + Q$$

$$(9)$$

The above reactions take place in theoretical conditions. However, practically, oxygen is not used entirely: some of it is wasted with exhaust gas. Therefore, what occurs is an incomplete combustion, connected with evaporation of the fuel and formation of its mixture with air, as mentioned above.

In practice, the quantity of oxygen introduced into the combustion chamber together with air is larger than its stoichiometrically indispensable quantity. The ratio between the actually introduced quantity of air and the stoichiometric (or theoretically required) quantity is called "excess-air ratio", α. For $\alpha < 1$, the fuel-air mixture is lean; for $\alpha = 1$, the mixture is stoichiometric, and for $\alpha > 1$, the mixture is rich.

For fuels which contain sulfur (which is the case in some diesel fuels and heating oils), an additional amount of oxygen is needed for oxidation of that element, and it is determined accordingly.

Reduction of the combustion product vapor volumes, as compared with the gaseous reactants, is called chemical contraction; when combined with physical contraction resulting from reduction of product volumes due to steam condensation, it is called "total contraction".

Depending on temperature range, fuel oxidation comprises three stages:

- slow oxidation at temperatures up to 200°C;
- low-temperature oxidation – slow, in gaseous phase at temperatures in the range (200…600)°C;
- fast oxidation in flame.

Although the first step has no major role in the engine technology, it does matter in determining the permissible duration of storage of fuels.

At low temperatures, particles have low kinetic energy so the number of fuel particles colliding with oxygen particles is small, therefore, oxidation process is very slow. At temperatures higher than 200°C some fractions in light fuels evaporate while the probability of collision between the evaporated fuel and oxygen particles is higher, which leads to the formation of low-temperature oxidation products (alcohols, aldehydes, ketones, acids and hydroxyacids in what is called "the first stage of oxidation", and peroxides, of which the structure comprises an oxygen bridge made of two oxygen atoms bonded together). Intermediate oxidation products have different kinetics of their further reactions with oxygen. The oxidation products mentioned above (except peroxides) have similar combustion kinetics. Their formation enhances the heat effect and the fuel combustion process rate, leading in effect to what is called "normal course of combustion" where the flame front moves away from the ignition source at a speed of (15...30) m/sec.

Fuels are multicomponent mixtures, in which paraffins are the hydrocarbon components having the lowest oxidation stabilities (the lower, the longer are their backbones and the lower are their isomerization degrees). Compared with paraffins, naphthenes have better oxidation stabilities because of their cyclic structure, and aromatic hydrocarbons have the highest oxidation stability. In the case of aromatic hydrocarbons with side chains, oxidation leads primarily to the breaking of these side chains whereby oxygen hydrocarbon derivatives are formed, after that the ring is broken and carbon oxides and water are generated.

In all the three hydrocarbon groups, during oxidation there may also appear intermediate combustion products which comprise an oxygen bridge-O-O-(peroxides). At lower oxidation temperatures, the duration of low-temperature oxidation is longer and, therefore, the amount of peroxides formed is higher. In high-temperature oxidation conditions, no peroxides were detected, because they are highly unstable and readily decompose. Hence, if low-temperature oxidation does not lead to the formation of peroxides, combustion is considered as a normal combustion reaction which proceeds gradually. Any intermediate oxidation products that may be formed and are in "first stage of oxidation" because of their higher stability, do not usually lead to the formation of peroxides.

The peroxides formed during low-temperature oxidation are highly reactive. The oxygen bridge in their structure may be present in the chain or form a bypass.

Having such structure, the compounds are very strong oxidants. Hydrogen peroxide is the simplest peroxide. Nitric acid(V) (HNO_3) is also a very strong oxidant. The content of active oxygen in their molecules is sufficient to initiate ignition or even combustion in the absence of atmospheric oxygen. Peroxides, which are formed during oxidation of hydrocarbons, being active particles, are able to initiate a combustion chain reaction. In such type of reactions, each active particle initiates another reaction, leading to new particles, initiating "n" subsequent reactions.

The combustion reaction accelerates violently till the process is complete. In that case, the whole fuel-air mixture is combusted violently in practically one moment. Such combustion process leads to very high values of temperature and pressure in the engine combustion chamber. It generates a shock wave which travels at a speed of (1500...2500) m/sec and, when colliding with the combustion chamber walls, causes the metallic knocking effect with sound frequencies of around 3000 Hz. This type of combustion is highly disadvantageous and is termed "detonation combustion".

Detonation combustion occurs around 0.6° crankshaft rotation (while the position of the flame front, developed by the initiated, ignition hardly changes at all). Interference of waves being created by the various detonation sites is observed: the waves amplify or dampen at the nodes. After such combustion, new mixture portions are introduced into the overheated combustion chamber during the consecutive intake cycles. Low-temperature oxidation is accelerated while the ignition delay remains the same. This increases the probability that new active particles will be formed and will initiate the detonations. Moreover, at too high temperatures, the fuel getting through in its liquid form is unable to evaporate fast enough and is thermally decomposed, which generates an immense amount of gaseous decomposition products and layers of carbon deposit on the walls (they have a thermally-insulating effect and reduce the capacity of the combustion chamber).

The sum of these phenomena very much improves the possibility that a significant number of detonation sites will soon develop. The subsequent strokes, which are performed at high speeds, can lead to the incomplete removal of combustion products from the chamber – these products have then enough time for being oxidized into the form of active particles. Therefore, detonation combustion can be intensified very rapidly and lead – in extreme cases – even to engine damage or destruction. However, not every case of active particle formation leads to a chain reaction or avalanche effect because the reaction tends to slow down as the peroxides develop. This is caused by the formation of final products of the combustion of carbon oxides and water as well as its intermediate products, such as alcohols, aldehydes etc. The combustion chamber walls also tend to deactivate the active particles being formed.

Group composition has a decisive influence on the nature of combustion of fuels because the respective hydrocarbon groups have different oxidation tendencies. Intensity of oxidation, resulting in the formation of active particles, decreases in the following order: alkanes (paraffins) > alkenes (olefins) > alkadienes (diolefins) > cycloalkanes (naphthenes) > arenes (aromatic hydrocarbons). Depending on their isomerization degree, higher branched isoparaffins show higher resistance to detonation combustion. Thus, e.g., 2, 3-dimethylpentene has a detonation combustion resistance similar to that of benzene, while that of 2, 2, 3-trimethylpentane is much higher. Apart from initial oxidation products, which are directly responsible for detonation combustion, there exist a number of direct and indirect factors increasing the possibility of peroxides being formed in the combustion chamber.

Indirect factors include:
- design factors;
- operating factors;
- physico-chemical properties of fuels;
- chemical composition of fuels.

Direct factors include
- temperature;
- pressure;
- duration of contact with oxygen.

Further, these factors can be grouped as follows:
- Design factors:
- compression ratio;
- pressure charging;
- size and shape of the combustion chamber;
- material of construction of the combustion chamber;
- quantity and distribution of sparking plugs.

Operating factors:

- ignition advance angle;
- composition of the fuel mixture;
- motor load;
- RPM;
- performance of the cooling system.

Compression ratio is one of the most important factors affecting the type of combustion. At what is called the "critical compression ratio", detonation combustion takes place due to an increase in pressure and temperature. The compression ratio (ε) is the ratio between the total capacity of the combustion chamber (defined as the sum of engine displacement and combustion chamber capacity) to the combustion chamber capacity.

In order to prevent detonation combustion at high compression ratios, it is necessary to use fuels with high oxidation stabilities. The value of critical compression ratio is also the measure of a given fuel resistance to detonation combustion.

An increase in the pressure of the air entering the combustion chamber, which is usually carried out by so-called pressure charging, increases the fuel's tendency for detonation combustion. This leads to higher temperatures during the compression stroke, thus intensifying oxidation of the fuel vapor which causes the formation of active particles.

The size and shape of the combustion chamber have an effect on the reaction rate, the amount of heat generated, and on the pressure increase; therefore, the fuel's tendency for detonation combustion increases with the cylinder capacity. The more complex the combustion chamber is, the greater the fuel's tendency for detonation because, after the flame front passage, there may remain some unburnt fuel particles which continue to react with oxygen and, at elevated temperatures, they may react, forming active forms which initiate detonation.

The probability that detonation sites will be formed is lower for less intense removal of heat from the combustion chamber; such intensity is provided by an efficient cooling system and good thermal conductivity of the chamber construction material. For example, an aluminum combustion chamber, due to the material's higher thermal conductivity, allows the use of higher compression ratios, compared with that of a cast iron chamber. A kind of catalytic effect of the construction metals on the course of initial fuel oxidation is observed, but the effect is not fully understood.

The probability that detonation sites will be formed is lower for smaller cylinder diameters and increased quantities of sparking plugs. Additional, properly distributed sparking plugs reduce the path length of the flames spreading from the ignition point to the end of the combustion chamber, thereby reducing the duration of contact between the fuel particles and oxygen. Gasoline-fuelled aircraft engines have two sparking plugs, arranged as follows: one near the intake valves and the other near the exhaust valves.

Design factors affecting the combustion process should be taken into account in engine designing and in selecting the proper fuel for them. But the operating factors, discussed below, which influence the course of fuel combustion ought to be adjusted accordingly during the operation of motors. Correct maintenance should help almost completely eliminate their adverse effect.

Thus, the higher the ignition advance angle, the higher (and correlated with an increase in the maximum combustion pressure) is the fuel's tendency for a detonation combustion. During early ignition, temperatures of the head and of the combustion chamber walls will rise; however, if the ignition is too late, the temperature growth is observed in the exhaust valve and the exhaust pipe. Hence, very often, when knocking is heard, normal combustion can be achieved by delaying ignition.

Another operating factor which influences the formation of peroxides is the fuel mixture's composition, the associated contamination of the mixture with exhaust gases, and the characteristics of the intake air. The most pronounced tendency is exhibited by mixtures of which the composition is similar to the stoichiometric one. This is connected with the highest burning rates being attained for excess-air ratios in the range of 0.95...1.05.

Throttling the suction process reduces the degree of filling, thus, it reduced the amount of heat being released during the process, which in turn reduces the fuel's tendency for detonation combustion. The fuel's tendency for detonation combustion increases at full engine loads and low rotational speeds. Throttling causes increased contamination of the mixture by exhaust gases which, by isolating the vapors with inert combustion products, increases resistance to detonation combustion, thus reducing the necessary activation energy in the particles.

The higher the atmospheric pressure, the higher is the probability of formation, in the pre-oxidation process, of active particles which initiate detonation combustion. If this happens, pressure at the end of the suction process will increase, resulting in a more accurate filling of the combustion chamber with the fuel mixture. The temperature rise due to the large mixture volume enhances oxidation in that portion of the mixture where combustion is not yet taking place.

In gasoline aircraft engines, the composition of the air varies with altitude so that the content of oxygen is lower and that of nitrogen is higher at higher altitudes. The mixture becomes leaner in oxygen, and the inert nitrogen has an isolating effect and reduces the activation energy, thus reducing the tendency for detonation combustion.

As the temperature of the air entering the fuel increases, so does its evaporation rate (and a more homogeneous mixture is obtained, although the fuel's tendency for detonation combustion is higher, too).

Higher air humidities lead to lower temperatures and pressures of the gas in the cylinder, while the dissociated steam (having better anti-knocking properties in comparison with inert gases) accelerates combustion, and a lesser amount of heat evolves from the wet mixture. The tendency for detonation combustion grows with the number of revolutions (RPM) because of the shorter residence time of the fuel in the high temperature zone, where it is pre-oxidized from the moment of ignition. The ability to form active particles is then reduced, the combustion process is accelerated and heat exchange between the fuel mixture and the heated walls of the combustion chamber is intensified.

The increased engine loads lead to higher temperatures which, in turn, cause detonation combustion.

The performance of the cooling system and the resulting intensity of the cooling process have a significant effect on the combustion process. Insufficient cooling may lead to oil burning on the cylinder bearing surface, seizing of piston rings, pre-ignition, and detonation combustion due to the temperature increase in the combustion chamber.

During combustion, thermal decomposition of the hydrocarbons may occur, causing settling of a carbon deposit on the combustion chamber walls which, in turn, leads to higher compression ratios and accelerated oxidation. Air cooled engines require fuels with a higher detonation resistance, compared with liquid-cooled engines. However, it should be kept in mind that engine overcooling leads to lower power outputs and has other consequences.

Physicochemical properties of fuels, such as degree of evaporation, volatility, boiling range, heat of evaporation, and chemical composition, have been discussed earlier in this chapter. From an analysis of the above, it follows that other indirect factors, for instance, temperature, pressure, and duration of contact with oxygen, also have an effect on the nature of combustion. However, it should be kept in mind that detonation combustion is directly caused by the formation of active particles having the structure of peroxides and acting as detonation combustion initiators. External signs of detonation combustion include a yellow flame and smoke in the exhaust pipe.

Engine operation is rough and jerky, with its output getting lower and lower. This causes rapid wear of the piston-crank assembly, valve burnout and deformation, burnout of piston heads, damage to gaskets and sparking plug insulations which, in intense detonation combustion conditions, lead altogether to engine destruction.

In some engines (particularly in aircraft engines), detonation combustion is prevented by injecting water or other liquids with low freezing points (such as ethyl-, methyl-, or isobutyl alcohols). The high heat of evaporation of the injected fluid reduces temperature of both the mixture and the combustion chamber walls, which impedes the detonation combustion. Moreover, steam being an inert gas interrupts the chain reaction and reduces the amount of carbon deposit. Another method to prevent detonation combustion is to use a mixture of water with the above mentioned alcohols.

Fractional Composition

In fractional composition determination by the so-called normal distillation method, the highest temperature obtained before the exhaust appear in the flask is assumed to be the final boiling point. It will drop slightly in the next steps of the process, then grow rapidly. The drop is observed because, after the light components have evaporated, the distillation residue is so heavy that the kinetic energy of the particles is too low to enable them to leave the liquid medium. Therefore, the liquid particles do not pass into vapor phase even though the flask is heated continuously and, therefore, no heat is carried to the thermometer. The rapid rise in temperature, as indicated by the thermometer, is caused by the thermal decomposition of the residue, leading to the formation of gaseous products of decomposition. In advanced measurement sets, the initial and final boiling points are determined by means of suitable, automatic sensors.

The starting capability of a fuel is assessed from the values of its initial boiling point and distillation 10% point. The lowest ambient temperature (to) which enables engine start is determined from the empirical relationship:

$$t_0 = \frac{1}{2}t_{10\%} - 50.5 + \frac{1}{3}(t_p - 50)$$

(10)

where:

$t_{10\%}$-distillation 10% point for gasoline;

t_o-initial boiling point.

Temperature $t_{10\%}$ should be a maximum of 80°C. If the initial boiling point is lower than 40°C, the risk of so-called vapor locks being formed in the supply system exists. The vapor locks, which consist of gasoline vapors having too high volatility, air bubbles, and heavier liquid components, tend to disturb the flow, leading to the fuel mixture rapidly becoming lean, and even to engine stoppages. The ambient temperature at which the vapor locks can be formed (t_o1), is found from the following relationship:

$$t_{o1} = 2t_{10\%} - 93$$

(11)

Because of the possibility of vapor locks formation, ambient temperature must not be higher than 76°C when the temperature under the hood is 60°C, and not more than 46°C for 0°C under the hood. Distillation 50% point for gasoline describes the fuel's average volatility and its average ability of evaporation and fuel mixture formation.

The values of distillation 90% and 97% points as well as the final boiling point characterize the fuels in terms of their content of heavy cuts. Such values should not be very high, otherwise, part of gasoline components would not be burned and the engine power output would be reduced for too-high fuel consumptions. Thermal decomposition of the fuel would be taking place in the combustion chamber, leading to overheating of the chamber walls, accumulation of too much coke and carbon deposits, resulting in an undesirable course of the fuel combustion process.

Vapor Pressure

Vapor pressure is the pressure exerted by the vapors of a liquid in a closed vessel onto its walls at a predetermined temperature. In hydrocarbon mixtures, such as gasoline, the parameter depends on the proportion of light components in the mixture. According to Dalton's law, the measured value of gasoline vapor pressure is the sum of the partial pressures of its individual components. It is not a constant value at a given temperature, since it depends on the concentration of its individual components in the test liquid and on the ratio between the volumes of the gaseous phase and the liquid phase. The higher the ratio, the lower the vapor pressure of the mixture.

The starting properties of gasoline are determined largely by its vapor pressure, which has a effect also on the formation of gas locks in the pipes (especially in aircraft engines at high altitudes). Therefore, saturated vapor pressure for aviation gasoline should be in the range from 0.03 MPa to 0.05 MPa.

The ambient temperature (t_o1) at which the gas locks can be formed, is found from the value of vapor pressure:

$$t_{o1} = 260 - 77.8 \log p$$

(12)

where:

p-vapor pressure of gasoline, as determined in Reid bomb.

The time it takes for the engine to start is dictated by volatility of gasoline and by ambient temperature. These relationships can also be illustrated using a suitable nomogram. These effects are also associated with fuel consumption, as shown in Table 1.

Table 1: Influence of ambient temperature on consumption of gasoline and engine starting time for a passenger car

A i r temperature [°C]	Engine starting time [sec] for gasoline		Fuel consumption [cm³] during engine starting time [sec] for gasoline	
	$t_{10\%} = 79°C$	$t_{10\%} = 72°C$	$t_{10\%} = 79°C$	$t_{10\%} = 72°C$
0	10.5	9.4	10.0	8,7
-6	45.0	29.0	480	70,0
-16	515.0	225.0	678,0	309,0

Resistance to Detonation Combustion

Critical compression ratio was the first criterion used for fuel assessment in the aspect of detonation combustion. The parameter was assumed to be the highest ratio at which combustion was normal. The value, however, depends on a number of subjective factors as well as on the engine condition and measurement conditions. So called "fuel equivalent methods" (using aniline, benzene or toluene) were also used. They compare combustion conditions for the test fuel solutions in reference gasoline with those of the same amount of fuel equivalent dissolved in the same kind of gasoline. The quantity of the fuel equivalent introduced indicated the fuel's resistance to detonation combustion. The study was carried out in a single-cylinder laboratory engine with variable compression ratio. However, the fuel equivalent methods were burdened with large errors associated with the resistance of the test solution (not pure fuel) to combustion detonation. In addition, the reference gasoline used showed different properties, depending on its origin and method of preparation.

In 1927, a criterion for the assessment of resistance to detonation combustion was developed and was adopted by the Standard Committees in most countries all over the world. The criterion is called the octane number or octane rating of fuels, and it measures their resistance to detonation. The octane number is determined using the octane number scale, based on the following, conventionally adopted, reference fuels:

- n-heptane (critical compression ratio=2.8; conventionally assumed octane number=0 units);
- 2, 2, 4-trimethylpentane – one of isooctanes (critical compression ratio=7.7; conventionally assumed octane number=100 units).

The adopted values are associated with the structure of the individual substances identified as reference fuels and with their resistance to detonation combustion. The adopted reference fuels have the physical properties of hydrocarbon components of gasoline.

The octane number is an absolute number, an integer equal to the percentage of 2, 2, 4-trimethylpentane in n-heptane in such a composition that the mixture, as prepared in the standardized motor in standardized conditions burns with the same detonation resistance as the test fuel.

To determine the octane number of fuels, the Cooperative Fuel Research designed and built a standard CFR engine. Other countries developed their own design, in accordance with the CFR engine requirements.

The test engines, designed for octane number determination, are single-cylinder, overhead-valve engines with a movable cylinder, cast as a whole with the head, having sensors (called "Midgley detonation needles") mounted in it, which enable evaluation of the intensity of detonation using the pressure-based method. As the cylinder is brought closer to or farther from the piston-connecting rod assembly through the worm gear system, the compression ratio of the engine during operation is changed.

A synchronous electric motor acts as a starter but also, after engine start, it receives the generated power by means of a belt drive. The reference engine is equipped with three fuel tanks and is fuelled through a carburetor having three float chambers which enable the engine to be supplied with a fuel mixture having different compositions of the reference fuels. Stable engine operating conditions are maintained by keeping stable temperatures of air, oil, and cooling water, and the optimum ignition advance angle for each compression ratio.

Depending on the measurement method, the engine can be provided with additional devices and measurements are made at the suitable rotational speeds. Conditions for octane number determination using the standardized motor according to various methods are listed in Table 2.

Only the so-called Army Method uses a different design of the combustion chamber. The method provides octane number values which are higher by 3 to 5 units, in comparison with the motor method. However, the Army Method has not become very popular and has been replaced by temperature-based methods.

Because of differences in engine rotational speeds between the research method (RM) and motor method (MM), some differences appear in the values of the octane numbers provided by the methods. RON is mainly used to determine the detonation combustion resistance of fuels, as used in passenger cars operated in the city or in field conditions at partial engine loads. RON shows generally higher values, compared with MON. The difference between RON and MON is defined as the fuel's sensitivity to the method of measurement. Out of two gasoline types having same MON but different RON values, the one with the higher RON is more sensitive, therefore, more useful. Similarly, among two gasoline types with same RON and different MON values, the one with the higher MON is more useful in operation.

Sensitivity of fuel as expressed by octane number values, can be as high as a dozen or so units. The lowest, sometimes negative, sensitivity is shown by paraffins (for example: pentane=-0.2, 3-methyl-hexane=-4.2). Higher sensitivities are shown by naphthenes (such as: ethyl cyclopentane=6.0, propyl cyclopentane=3.1) and the highest sensitivities characterize olefins (such as: butane-1=16.0, pentene-1=13.8) and aromatic hydrocarbons (toluene=12.0, propylbenzene=9.8).

Although olefins are highly resistant to detonation combustion, their presence in gasoline is not preferred because they tend to form resins during storage and sludge during fuel combustion in engines. However, for reasons of economy, they are used as additives to gasoline in amounts of about 30%.

MON is most commonly used for assessing the behavior of fuels in vehicles operated on longer routes (when the engine is exposed to more strenuous operating conditions, works in stable high-load conditions, at higher speeds and higher temperatures). MON is also used to evaluate resistance to detonation in less strenuous operating conditions for aircraft engines.

Test-stand methods for octane number assessment were found inadequate for determination of desired octane number values of fuels used in newly-manufactured cars with modified engines. Therefore, another method was developed for measurement of DON (roaD Octane Number). DON is determined in accordance with Uniontown and Borderline methods. In both of them, the fuel's behavior is tested while driving a car equipped with additional accessories on a road section of 1 km in standard conditions (such as surface, rectilinearity, wind power, etc.). The test car is provided with at least five tanks filled with fuels having different known RON values and is equipped with a device enabling measurement of the intensity of detonation, amount of fuel charge, linear velocity, rotational velocity, as well as measurement and change of the ignition advance angle.

In the Uniontown procedure, for each subsequent run 2° before TDC, every 2°, the car is made to run at a constant ignition advance angle until weak knocking for the given gasoline is observed. For the huge population of vehicles for which the test is carried out, it is possible to determine the ignition advance angle for the particular gasoline. The DON value as found by this method is usually lower than that of RON. The smaller the difference, the better the fuel's resistance to detonation combustion in operating conditions.

In contrast, the Borderline method estimates engine's rotational speed at which detonation ceases for a given fuel and for a given value of ignition advance angle. In the subsequent starting runs of a car, for gasoline having a known RON, the velocity is found for which detonation ceases at a pre-determined ignition advance angle which is varied in accordance with the characteristics recommended by the manufacturer. The detonation combustion region comprises part of the plot area limited by the characteristics of the ignition advance angle controller and part of sections in a pencil of curves.

Table 2: Operating conditions for a reference engine in determination of octane numbers by different methods

Parameters of motor operation	Method:					
	Research method (automotive gasoline)	Motor method (automotive and aviation gasoline)	ArmyMethod	Temperature method 1-C (aviation gasoline)	Temperature method 3-C (aviation gasoline)	
	RPM (revolutions per minute)	600	900	1200	1200	1800
Coolant type	water	water	ethylene glycol	ethylene glycol	ethylene glycol	
Coolant temperature [°C]	100	100	160	190	190	

Air temperature [°C]	-	38	-	52	107	
Oil temperature [°C]	55	55	70 - 80	65	74	
Mixture temperature [°C]	-	150	-	107	-	
Ignition advance angle (degrees before TDC)	26	22.6	30	35	45	
Cylinder diameter [mm]	82.6	82.6	66.7	82.6	82.6	
Piston stroke [mm]	114.0	114.0	112.0	114.0	114.0	
Detonation/ knocking indicator	pressure	pressure	Thermo-element	Thermo-element	-	

The approximate value of DON can be determined also from the octane index, which is the arithmetic mean of RON and MON.

For aviation gasoline with octane values of more than 100 units, ON measurements are carried out by 1-C and 3-C temperature methods. The methods are based on the cylinder wall heating-up with the higher intensity, the more intense detonation combustion occurs. A thermocouple is placed inside the cylinder head of the CFR engine with an aluminum piston, to determine the temperature rise gradient. The detonation temperature calibration line is found before starting the measurements. The composition of the fuel-air mixture is selected so that the thermocouple shows the highest temperature, and the compression ratio is selected in a manner that enabled the maximum temperature to be found in the reference temperature line.

In the case of aircraft engines with direct fuel injection and air-compressor turbocharging, the 1-C method is insufficient for determination of the correct octane number. Therefore, the 3-C temperature-based method was developed, whereby the octane number measurements are carried out by means of a seriously modified reference engine which is equipped with a compressor and an apparatus for measuring power outputs, fuel consumption, and air flows. Determination of fuel resistance to detonation combustion is based on finding the relationship between the mean indicated pressure (proportional to power output) and the mixture composition for the engine running with a slight detonation. Leading to detonation combustion is effected by increasing the supercharging pressure. The measurement starts with a lean mixture, and then the mixture is gradually enriched, the indicated pressure increases to a maximum and then drops. The plotted curve for mean indicated pressure vs. mixture composition is applied against comparable curves obtained for the reference fuels, thus determining the value of what is so-called "performance number".

Empirically, the performance number can be determined from the indicated pressure (Pi) according to the formula:

$$LW = 276 - \frac{29650}{P_i}$$

(13)

The quality of gasoline at the time of engine start and engine acceleration are characterized by the fuel's octane segregation index (R_{100} or R_{75}). Segregation index is the difference between RON for the test gasoline and for the fraction of the same gasoline boiling to 100°C (ΔR_{100}) or evaporating to 75% by volume (ΔR_{75}). The smaller the value of the segregation octane index, the more homogeneous is the gasoline. Such gasoline shows stable resistance to detonation combustion regardless of how much of it has evaporated.

The phlegm which is formed in the process of gasoline evaporation may lead to a non-homogeneous fuel (with different fractional compositions) being distributed into the cylinders and this will result in different resistances of the fuel to detonation combustion. The phenomenon is assessed by means of FON ("front octane number", formerly "distribution octane number"), measured by means of a reference engine, equipped with device called dephlegmator.

Dephlegmator is a water cooler for the flowing fuel vapors, with the cooling water temperature of 4.4°C. It is mounted between the carburetor and the inlet of the cylinder. The temperature of the mixture from the dephlegmator is in the range 17 to 22°C. Heavier components of the fuel are condensed in the dephlegmator fall to the bottom and are then sent to the measuring vessel; part of them (5... 7)% (V/V) is entrained into the combustion chamber. The fed mixture is slightly richer during the measurements for FON, compared with RON. FON will be much lower than RON in the case of phlegm formation in winter operating conditions in engines with carburetors (especially during the cold start).

Gasoline is a hydrocarbon mixture of about 100 different compounds obtained by straight-run or destructive processing. Table 3 summarizes the basic group compositions of most light distillate fuels.

With regard to the desired combustion method, gasoline should preferably have a relatively high content of aromatic hydrocarbons, most of which are obtained by reforming while some by pyrolysis. Very important components, especially in aviation gasoline, are mixtures of hydrocarbons produced in isomerization and alkylation processes, also called "isoparaffins". They have the advantage of high knock resistance, in addition to high calorific value.

Table 3: Group composition of light distillates

Hydrocarbons	Limits [%(V/V)]
Paraffins	40...65
Naphthenes	20...35
Aromatics	8...20

All the necessary ingredients for gasoline blending are obtained from different refinery processes and then handled in the blending unit, in accordance with the respective technological requirements for a particular gasoline grade, as discussed earlier in Chapter 1.

Ignition Properties of Fuels (Flash-Point, Self-Ignition Temperature and Fire-Point)

The respective fuel ignition limits depend on many factors, including: chemical composition of the fuel, temperature, and pressure of the mixture, ignition sites, etc. The problem applies to all motor fuels, from the engine gasoline, diesel fuel, heating oil, aviation fuel and biofuel. The essence of changes in these properties and, therefore, the criterion for measurement and the assessment of its weight and measure, depend mainly on the group-and structural composition for a given fuel. Therefore, general considerations about these properties are provided in the section concerning the first group of fuels, that is, gasoline.

The fuel mixture can be ignited when its composition falls within flammability limits and the temperature, called flash-point, is high enough (vapors are formed above the liquid surface at a concentration above the lower flammability limit).

Flash-point depends on the volatility of the liquid, in the first place. Light liquids with low densities (e.g., gasoline) have low flash-points although the value rises for liquids with higher densities. Flash-points of fuels are measured mainly by two types of methods: open cup – Marcusson, Cleveland or Brenken method or, depending on the type of fuel, using the closed cup: as in Abel-Pensky or Pensky-Martens method.

In closed cups, within the vapor space enclosed by the structure, the resulting mixture is unable to interact with the atmosphere so it will reach and exceed its lower flammability limit sooner (at a lower temperature) and ignite, if provided with the appropriate ignition energy by the initiator (such as a small gas burner). That is why closed cup testers are used mostly for determination of flash-points for light fuels.

Flash-point values obtained for a same fuel by a closed-cup method will be lower than those obtained by open-cup methods. Heavier fuels, having lower diffusion coefficients, in conditions of unlimited contact of vapors with the atmosphere above the liquid surface will make a suitable rich mixture much easier.

In the Marcusson method, correction for the flash-point reading is required because of the difference between the ambient pressure at which the measurement was carried out and normal pressure.

In the Marcusson and Brenken method if, after an initiated successful ignition, the fuel continues to burn for at least 5 seconds then the temperature of the heated fuel is called the fire point. At that temperature, the amount of heat supplied to the fuel is so large that, in the process of burning of the fuel vapor above the liquid surface, a sufficiently large number of particles evaporate therefrom to sustain the condition of the resulting mixture above its lower flammability limit. Therefore, fire point is higher than flash-point, at which the accumulated vapors will burn but the flame will not be sustained.

Self-ignition temperatures of fuels are higher than their fire points. In laboratory conditions, the values of self-ignition temperature of fuels are determined by the dynamic method, by dripping fuel droplets from the burette into the airstream flowing by a heated chamber.

The higher the fuel's density, the lower its self-ignition temperature. Self-ignition of fuels is associated with what is called their low-temperature oxidation, which is intensive for more complex particles with lower stability.

The qualitative requirements and test methods applicable to gasoline in the EU ought to be conformable with the said standard EN 228 in which fuels are categorized into volatility grades. The currently applicable division of gasoline into volatility grades in the EU countries is shown in Table 4.

Table 4: Volatility grades of gasoline

Properties	Units	Range						Test method[a]
		grade A	grade B	grade C/C1	grade D/D1	grade E/E1	grade F/F1	
Vapor pressure (VP)	kPa, min kPa, max	45.0 60.0	45.0 70.0	50.0 80.0	60.0 90.0	65.0 95.0	70.0 100.0	EN 13016-1[b]
% of evaporation to 70°C, E70	% (V/V), min % (V/V), max	20.0 48.0	20.0 48.0	22.0 50.0	22.0 50.0	22.0 50.0	22.0 50.0	EN ISO 3405
% of evaporation to 100°C, E100	% (V/V), min % (V/V), max	46.0 71.0	46.0 71.0	46.0 71.0	46.0 71.0	46.0 71.0	46.0 71.0	EN ISO 3405
% of evaporation to 150°C, E150	% (V/V), min	75.0	75.0	75.0	75.0	75.0	75.0	EN ISO 3405
Final boiling point (FBP)	°C, max	210	210	210	210	210	210	EN ISO 3405
Distillation residue	% (V/V), max	2	2	2	2	2	2	EN ISO 3405
Vapor Lock Index (VLI) (10VP+E70)	index, max	-	-	C-	D-	E-	F-	
Vapor Lock Index (VLI) (10VP+E70)	index, max			C1 1050	D1 1150	E1 1200	F1 1250	

NOTE: The requirements printed in bold type refer to the European Directive 98/70/EC, as amended by 2003/17/EC.

a)   See also 5.8.1.

b   The value of dry vapor pressure equivalent (DVPE) should be provided.

It follows, from the above considerations, that the following types of analysis are required in the assessment and maintenance of the quality of gasoline during storage:

Full analysis for conformity with the applicable standard specifications – to be performed when accepting the product for storage.

1.   Short analysis – to be performed at least once in 24 months, to assess the following parameters:

- appearance;
- density;
- vapor pressure;
- fractional composition;
- content of inherent resins;
- optionally, content of oxygen compounds.

2.   Control analysis – to be performed at least once in 6 months, to assess the following parameters:

- appearance (visually);
- water content (visually);
- impurities (visually);
- density.

According to NATO standards, the permissible safe time of storage of gasoline is 36 months.

If the short or control analysis of the fuel parameters indicates limit values for the fuels, as shown in the applicable standard specifications, or the results are dangerously close to the limit values, then a full analysis is required and its findings will show how to deal with the stored fuel.

A control analysis ought to be carried out after any technical procedure was performed with regard to the storage tank with the fuel in it, and a short analysis is applicable after any procedure was performed with regard to the fuel being stored in the tank.

Whenever the whole fuel batch is released, the short analysis should essentially be sufficient subject to acceptance from the recipient but even so, full analysis is the recommended solution to prevent any complaints.

CRITERIA FOR THE ASSESSMENT OF THE QUALITY OF DIESEL FUELS IN STORAGE AND OPERATING CONDITIONS

Diesel fuels are mixtures of liquid hydrocarbons having boiling ranges (180...380)°C, with a content of improvers and they are used as fuels for engines with spontaneous-ignition (diesel, or self-ignition engines). Diesel fuels are obtained by blending the appropriate products of crude oil distillation and other refinery processes.

Differences in the diesel fuels of to-day are found, first of all, in their having different cold filter plugging points (CFPP), freezing points, and viscosities. Many users still tend to believe that diesel fuels are a residual, poor quality product. Generally though rather wrongly, diesel fuels are regarded by such users as a deteriorated fuel with low requirements concerning quality, production methods, storage, and performance in operating conditions. Another prevailing conviction is that most diesel engines can be fuelled with poor-quality fuels, an example of which diesel fuels are believed to be. However, all the requirements relating to advanced engine designs, to the complex advanced methods for the obtaining of diesel fuels, as well as the ever more stringent environmental protection requirements are ignored in this approach.

Existing diesel fuels are fuels with high qualitative requirements and sophisticated testing methodology. This results from certain radical changes in the design of self-ignition engines, as well as more stringent requirements concerning the various consumable fluids used in such engines. Not only are such requirements rather essentially connected with the design of engine fuel distribution and combustion systems but also with environmental impact issues and severe control of harmful emissions.

Low-Temperature Properties of Diesel Fuels

Rheological properties of diesel fuels, especially in low-temperature conditions, have a significant impact on the performance of fuel and combustion systems in diesel engines. They determine the possibility of delivering sufficient amounts of oil into the combustion chamber. This is especially important when starting the engine at low ambient temperatures because crystallization of some of the components of diesel fuels may occur in such conditions. Paraffin hydrocarbons (waxes), otherwise a desirable component of diesel fuels because of their self-ignition properties, tend to crystallize at the highest temperatures.

Crystallization of hydrocarbons in the fuel leads to filter plugging, which causes significant flow resistance, thus reducing fuel delivery. Intensified crystallization may cause complete plugging and stop the fuel from flowing through the filter: this occurs after a more than 3 mm thick deposit of crystals has accumulated on the filter. The temperature at which the phenomenon occurs is called the cold filter plugging point (CFPP), or critical filterability temperature, and is a very important criterion in analyses intended to check the quality of diesel fuels in the aspect of approval for use.

CFPP is determined by the Hagenmann and Hammerich method by measuring the highest temperature at which a diesel fuel stops flowing through a standard filtration system in standardized conditions, or the flow of 20 cm^3 of the diesel fuel takes more than 60 sec. CFPP is about (10...15)°C higher than the freeze point of a diesel fuel, defined as the temperature at which the mobility of the diesel fuel sample is reduced so that its meniscus in a tilted test tube at an angle of 45° does not move within less than 60 sec. The cloud point of diesel fuels, as defined in standard specifications, is above the freeze point. The value of crystallization point is also stated in some sources. Both these temperature values characterize the same physical phenomenon, though it is visualized in different ways. When measuring the cloud point, the standardized sample becomes cloudy; in crystallization temperature measurements, single crystals which are visible to the naked eye start forming in the sample as it is cooled down.

The above values, especially CFPP, are very much affected by the sample's contamination, water content, and duration of storage in storage tanks, in addition to fractional composition. Mechanical impurities (solids) will deposit on the filters, thus accelerating filter plugging. In addition, dust microparticles and impurities tend to attract wax particles and favor their crystallization. While the content of water has a highly undesirable effect on the low-temperature properties of diesel fuels, during their long-term storage, resinous ageing products may form in them, which disturb their flow through the filter and accelerate crystallization.

According to European requirements for the temperate climate for the various diesel fuel grades, the following CFPP values are permissible: from 5°C max. for grade "A"; followed by 0,-5,-10,-15°C for the consecutive grades, with-20°C for grade "F". For the so-called arctic oils category, the CFPP values for grades 0 to 4 are-20,-26,-38,-44°C, respectively. The cloud point values "tolerated" in applicable standards are 10°C higher for each oil grade.

Table 5 shows applicable requirements, according to the European standard, concerning the most important low-temperature properties of diesel fuels, intended for use in so-called arctic climate conditions.

Table 5: Low-temperature properties of diesel fuels intended for use in arctic climate conditions, according to EN 590:2013

Property	Range					Test method
Oil grade	0	1	2	3	4	-
Freezing point in °C max.	-10	-16	-22	-28	-34	EN 23015
CFPP in °C max.	-20	-26	32	-38	-44	EN 116

Requirements Connected With the Evaporation of Diesel Fuels

Evaporation of diesel fuel in a self-ignition engine, as a process, has essentially two phases.

The first phase takes place from the start of fuel injection to the time self-ignition; it takes place at the cost of the heat contained in the compressed-air combustion chamber. During that phase, the evaporating hydrocarbons are preliminarily oxidized while evaporation is accelerated and intensified by the heat of exothermal oxidation reactions. This step is also called self-ignition delay and, as explained later in this chapter, it occurs at the same time as the first step of combustion.

The second phase of evaporation takes place from the time of self-ignition to the completion of fuel injection. Evaporation of the fuel, in this case, takes place at the same time as combustion, at the cost of the heat being generated during the combustion of the previously evaporated portions of fuel.

In engines equipped with pre-combustion chambers, evaporation starts at the time the fuel is injected into the pre-combustion chamber. The rich mixture formed in it has a composition which enables self-ignition. This occurs at sufficiently high temperatures; part of the mixture is combusted. The rapid evaporation of a portion of non-evaporated fuel causes a pressure increase, whereby the fuel during combustion is forced into the main combustion chamber and the fuel combustion process is continued.

The fuel's delivery and evaporation is much affected by the viscosity of diesel fuel. At 20°C, the viscosity of diesel fuels for high-speed diesel engines is typically between (2.8...8.0) mm²/sec.

For every type of engine, there is a limit to the fuel viscosity which, for a given design of the fuel system, makes impossible normal power delivery because flow resistance is too high, leading to flow disturbances. Viscosity of diesel fuels decreases as the temperature increases, this affects the conditions of flow. The least observable changes in viscosity, with temperature variations, are shown by the paraffin fractions of diesel fuels. If present in excess in the oil, the paraffin fractions will improve its spontaneous-ignition properties but its low-temperature properties will be much worse (higher cloud point and CFPP values).

With a decrease in the oil viscosity, the fuel stream's atomization and evaporation are lower though the fuel stream can reach farther, therefore, the fuel tends to settle on piston heads and chamber walls, leading to the formation of carbon deposits. Too low viscosities will affect lubrication of the injection pump pistons and reach of the fuel stream, vary the distribution of the fuel droplets in the combustion chamber, lead to incomplete combustion and to the presence of local hot spots in locally overheated walls of the combustion chamber. Low viscosities may result in fuel leakages from precision pairs of components and in reduced fuel dosage.

A fuel stream, delivered by the injection pump through the injectors, is composed of several million droplets of fuel, the size between (3...5) μm and (100...150) μm. The fuel spraying quality is characterized by its droplet size and quantity or, to be more precise, by the degree and uniformity of spraying, the range of the fuel stream, and angle of spraying cone.

The degree of spraying is defined as the average size of the fuel droplets going out of the injector. Stream uniformity is to be understood as the ratio between the number of average-size fuel droplets to the total number of fuel droplets.

The quality of spraying depends on the following properties of diesel fuels, in addition to the injector nozzle design:

- viscosity (discussed earlier);
- density;
- fractional composition;
- vapor pressure;
- surface tension;
- heat of evaporation;
- heat of combustion.

Evaporation of diesel fuel is easier at higher values of spraying degree and uniformity. Like the fuel's viscosity, its density affects mainly the range of the fuel stream in the combustion chamber. Moreover, lower densities of diesel fuel lead to reduced emissions of solids to a linear course abut also, in certain instances, may lead to reduced emissions of NOx. Lower density of diesel fuel is connected with its lower calorific value – this is of importance to engine performance. In this case, inhibition of engine output reduction by increasing the fuel dosage leads to higher fuel consumption levels, eliminating the solids emission-reducing effect. Reduction of density may also result in a degree of reduction of CO_2 emissions by a maximum of 1%. Therefore, the density of diesel fuels should be suitably low and vary over a small range only. According to applicable European standards, the permissible density of diesel fuels is (820...860) kg/m^3 at a temperature of 15°C, though its upper limit is expected to be lowered to 845 kg/m^3.

Uniformity of fuel spraying is proportional to its surface tension which, in turn, depends on the presence in the oil molecules of polar links. Along with the increase in fuel density, surface tension grows and, at the air-oil interface, is in the range $(27 \bullet 10^{-7}...30 \bullet 10^{-7})$ J/cm^2 and getting lower as the temperature increases: this is favorable for fuel spraying. Paraffin-naphthene fractions in the fuels have much lower surface tensions, compared with aromatic hydrocarbons.

Fuel evaporation depends also on its fractional composition which is determined by means of normal distillation. The drip point has an effect on the engine start characteristics. The distillation 50% point correlates with the viscosity and density of fuels, affecting the degree and uniformity of spraying, as well as on the stable course of evaporation and combustion processes and on the easy engine start. The distillation 90 and 95% points as well as the final boiling point have a significant impact on toxic emissions. According to the applicable EU Directive, the maximum distillation 95% point for diesel fuels is 370°C, and is to be lowered to 360°C. If the evaporation of light fuel fractions is fast enough, then the time required for making a homogeneous combustible mixture is short. On the other hand, there takes place a rapid pressure build-up, leading – after self-ignition of the mixture – to rough engine operation, which is undesirable. Heavy fractions in diesel fuels may lead to the incomplete combustion of the fuel, causing thermal decomposition of non-evaporated fuel droplets, which is accompanied by the formation of large amounts of soot in the exhaust gases and carbon deposits on injector tips. Moreover, the non-combusted fuel flowing down the combustion chamber walls may wash down the lubricating oil, thus accelerating the wear and tear of cylinder sleeves.

The heat of evaporation of fuels, though not determined in applicable standards is taken into account in the processes of collection of diesel fuel cuts from the distillation column. The value of heat of evaporation determines how readily a combustible mixture is formed in cold engine start conditions. As stated earlier in this chapter, such properties may also be determined by means of drip point and distillation 50% point for diesel fuel.

According to European requirements, the volumes of distilled diesel fuel are established for operation in arctic climate conditions for temperatures 180°C and 340°C, as shown in Table 6.

Table 6: Properties of diesel fuels for use in arctic climate conditions, connected with evaporation, according to EN 590:2013

Property	Range						Test method
Oil grade	0	1	2	3		4	-
Density at 15 °C in kg/^{m}3	800...845		800...840				EN ISO 3675 EN ISO 12185
Viscosity at 40 °C in mm²/sec	1.50...4.00			1,40...4.00		1.20...4.00	EN ISO 3104
Fractional composition: %(v/v) distill. to 180°C max. %(v/v) distill. to 340°C min.	10 95						ISO 3405

Requirements Connected With Diesel Fuel Combustion

Diesel fuel combustion, which takes place in self-ignition engines, is a three-phase process.

Phase 1 is the time of self-ignition delay, in other words, the time of preoxidation of the fuel. It starts at the time of fuel injection, at (20...30)° of crankshaft rotation before TDC (Top Dead Center) and takes about 0.0007sec, that is, until self-ignition. Phase 1 overlaps entirely with the first step of evaporation of fuel. If the fuel preoxidation is more intensified (which happens at higher temperatures and pressures in the combustion chamber), then the duration of the self-ignition delay is shorter and the engine operation is smoother. The duration of the first phase of combustion depends predominantly on the fuel's chemical composition.

Phase 2 of the process is the actual combustion of the fuel that was accumulated and prepared for combustion in phase 1. What occurs here is a fast combustion and intensified pressure buildup in the combustion chamber. A flame appears in many spots in the combustion chamber; it originates from the respective self-ignition sites, created by active radicals. The duration of phase 2 depends on the evaporated amount of fuel and the homogeneity of its mixtures with air. A rapid pressure buildup in the second phase of combustion may generate effects which are comparable to a detonation combustion in spark-ignition engines. Its duration determines the operation of a compression-ignition engine. With short self-ignition delays, pressure increments are rather mild and the engine operation is smooth. Long-lasting self-ignition delays may lead to a number of undesirable phenomena.

Rapid pressure and temperature increments lead to the formation of too much carbon deposit due to thermal decomposition of the fuel. Moreover, loads acting upon the piston, connecting rod and bearings can be too high, engine output and its efficiency are low. Exhaust gases are contaminated by soot and toxic fuel decomposition products, smoke level is higher and metallic knocking is heard in the cylinders. The engine operation in such conditions is termed "rough".

Phase 3 is a step of delayed, controlled combustion during which the fuel rapidly evaporates and is combusted as it is injected until the injection process is complete. For normal engine operation, settings and fuel choice, the pressure increase rate is not expected to be higher than approx. 588 kPa/sec.

Among the components of diesel fuels, the best self-ignition properties are shown by long-chain paraffins. The isomerization degree for those hydrocarbons determines for the duration of the self-ignition delay. Unsaturated hydrocarbons, or alkenes, have self-ignition delays close to those of their respective paraffins and isoparaffins. Naphthenes have better oxidation stabilities, compared with paraffins. The side-chain in paraffins usually extends the self-ignition delay in naphthenes.

The longest self-ignition delays are observed in aromatic hydrocarbons: they are proportional to the number of rings per molecule. The presence of a side chain reduces the self-ignition delay for aromatic hydrocarbons, especially in the case of unbranched chains.

Generally, the higher the boiling point for the hydrocarbons in diesel fuels, the better their self-ignition properties. The self-ignition tendency of diesel fuels is defined in standard specifications by means of cetane number and cetane index.

The cetane number (CN) of diesel fuels, which measures their self-ignition tendency, is defined as a non-denominated integer which expresses the percentage by volume of the standard fuel n-cetane $(C_{16}H_{34})$ of which the cetane number is assumed to be 100 units, contained in a mixture with the standard fuel α-methylnaphthalene of which the cetane number is assumed to be 0 units, so that the resulting mixture in standard conditions in a standard engine is combusted showing the same self-ignition tendency as the test fuel.

The measurement of CN is a complex procedure which requires special test engines, depending on their settings, and is carried out by delayed self-ignition methods, when the fuel injection is set at 10° of crankshaft rotation before TDC and self-ignition is effected by varying the compression ratio, 1° of crankshaft rotation after TDC, or by synchronization of fuel injection, by analogy: fuel injection 13° of crankshaft rotation before TDC, self-ignition at TDC.

Too low CN values of less than 45 units lead to deteriorated engine operation and the self-ignition delay is too long. This may cause excessive temperature increase and pressure buildup in the combustion chamber, leading to rough engine operation which causes a premature wear and tear of engine components and makes difficult the cold engine start.

The combustion of fuels with CN values of more than 70 units will also lead to deteriorated course of the combustion process because the fuel combustion is incomplete and the self-ignition delay is very short, causing too high smoke levels in the exhaust gases and inefficient engine operation.

According to the applicable European requirements, the permissible minimum value of CN is 51 units. For fuel grades intended for use in arctic climate conditions, the minimum CN values, according to the same European standard, are somewhat lower, especially for higher fuel grade numbers.

Because the CN measurement procedure is so complex, the notion of cetane index (CI, CCI) has been introduced in standard specifications for diesel fuels. It characterizes the self-ignition properties of diesel fuels as well, although it can be calculated from their densities and the course of their normal distillation.

According to ASTM D976-80, the value of CI is calculated as follows:

$$CI=454.74-1641.416D+774D^2-0.554B+97.803(\lg B)^2$$

(14)

where:

D – fuel density at a temperature of 15°C;

B – the temperature at which 50% vol. of the fuel has distilled (distillation 50 point, °C).

Another method to obtain the Calculated Cetane Index (CCI), in accordance with ASTM D4737-87, is more complicated and requires the knowledge of the fuel density and distillation 10, 50 and 90% points of fuel: CCI is calculated from the following relationship:

$$CCI=45.2+(0.0892)(T_{10N})+[0.13]+(0.901)(B)(T_{50N})+[0.0523-(0.0420)(B)][T_{90N}]+$$
$$+[0.00049][(T_{10N})^2-(T_{90N})^2]+(107)(B)+(60)(B)^2$$

(15)

where:

$B=[e^{(-3.5)(DN)}]-1$;

$DN=D-0.85$;

D – fuel density at 15°C;

T_{10} – the temperature at which 10% vol. of the fuel has distilled (distillation 10 point, °C);

$T_{10N}=T_{10}-215$;

T_{50} – the temperature at which 50% vol. of the fuel has distilled (distillation 50 point, °C)

$T_{50N}=T_{50}-260$;

T_{90} – the temperature at which 90% vol. of the fuel has distilled (distillation 90 point, °C);

$T_{90N}=T_{90}-310$.

The latter method to calculate the cetane index (CCI) is currently adopted as applicable in standards. The difference between the cetane index and cetane number values is 1...4 units, and the former is usually lower. The minimum value of cetane index for diesel fuels intended for use in temperate climate conditions is set to be 46 units.

The flash point value for diesel fuels is also established in standard specifications. According to European requirements, its minimum value is higher than 55°C; the value applies to diesel fuels for both the temperate and arctic climate conditions (Table 7).

Table 7: Properties of diesel fuels intended for use in arctic climate conditions, relating to the combustion process, according to EN 590:2013

Property	Range					Test method
Oil grade	0	1	2	3	4	-
Flash point °C	"/>55					EN 22719
Cetane number, min.	47.0			45.0		EN ISO 5165
Cetane index, min.	46.0			43.0		EN ISO 4264

Environmental Impact Requirements for Diesel Fuels

The presence of sulfur in diesel fuels and their aromatic hydrocarbon content is extremely important, in addition to the above-mentioned correlations between the normative properties of diesel fuels and the environmental impact of such fuels.

Sulfur and its compounds – which are hard to remove from diesel fuel during its production process – may have a direct corrosive effect on engine construction materials, as shown by the following reaction:

$n(R\text{-}SH)+Me \rightarrow (R\text{-}S)_n+nH_2;$

$n(H_2S)+Me \rightarrow n(MeS)+nH_2;$

$nS+Me \rightarrow n(MeS).$

The corrosive effect of elementary sulfur is observed at as low concentrations as 1mg per 100 ml of fuel.

During the diesel fuel combustion process, their sulfur content is oxidized to form sulfur trioxide and dioxide which combine with steam forming, respectively, solutions of sulfuric(VI) acid and sulfuric(IV) acid, which – owing to their presence in the exhaust gases – have a strong corrosive and acidifying effect on the environment. A certain amount of sulfur (1..2) %(m/m) reacts to form insoluble sulfates which are emitted in the form of solids with the exhaust. Their emissions may be high in the case of vehicles equipped with oxidation catalysts. Moreover, sulfur has a destructive effect on reduction catalysts, which may result in worse NO_x conversions into N_2.

In addition to their adverse effect on the self-ignition properties of diesel fuels, aromatic hydrocarbons lead to increased emissions of solids, lower reduction of NOx emissions, and direct pollution by toxic emissions of cancerogenic hydrocarbons, especially multi-ring aromatic hydrocarbons.

Other Performance Requirements for Diesel Fuels

There are a category of properties of diesel fuels (and they are characterized in applicable standards), which have an indirect effect on all of their performance parameters: on the course of evaporation and combustion but also on environmental hazard. Such properties include the following;

- water content;
- solids content;
- ash residue after incineration;
- carbon residue in 10% distillation residue;
- oxidation stability;
- lubricity.

Lubricity was implemented in EN 590 in the year 1998. The requirement of its determination for diesel fuels is justified by the need to prevent any unnecessary wear and tear, particularly with regard to precision pairs of components in the fuel distribution system. This is especially important in view of the low-sulfur content standards which are applicable to diesel fuels, because some sulfur compounds appear to have good wear and tear properties. Advanced compression-ignition engines are characterized by low tolerance, very high machining precision of components, and their structural materials are frequently modified.

The effect of the other above-listed properties of diesel fuels on the essential engine operation parameters and on the toxicity of exhaust gases was discussed earlier in this chapter. Water content and solids disturb the process of fuel transport and distribution in the fuel system; at near 0°C temperatures, they favor clouding (crystallization) and freezing processes and increase the value of CFPP.

Oxidation stability is a property which may determine the permissible duration of storage of diesel fuels, and may affect the course of combustion of such fuels. Very high values of oxidation stability will extend the permissible duration of storage of diesel fuels but they also affect combustion, for instance, by extending the self-ignition delay; this may lead to rough engine operation and its further undesirable consequences. There are good reasons to believe that the value of oxidation stability according to the applicable European standard is not essentially going to change in the near future, unless the standard related to the test method has been amended.

The coking number of diesel fuels indicates their tendency to form carbon deposits and is found from 10% of distillation residue. The phenomenon may change the thermal conditions prevailing in the combustion chamber, and lead to a deteriorated course of the combustion process, formation of local temperature gradients and, consequently, also high stresses in construction materials. The phenomenon, combined with the value of ash residue after incineration, may indicate the possible emission of solids with the exhaust gases. According to the requirements of the respective environmental protection agencies, such emissions are going to be more and more strictly limited. Emission level depends on the type and amount of improvers being added to diesel fuels. The limit on the coking number, as set out in the applicable European standard, ought to be maintained by fuel manufacturers before they choose to use of any additives that are known to increase the value of CN. If such an additive is detected in a commercial fuel by means of a test referred to in EN ISO 13759 concerning the presence of nitrate additives, then the limiting coking number as set out in that standard is not applicable.

As regards the essential properties of diesel fuels, in the assessment of their quality and performance in accordance with applicable standards, it is not necessary to determine their foaming tendency. Silicone-based foam inhibitors are used in diesel fuels in many countries except for the USA, where the material has not been approved by the EPA. Antifoaming properties of diesel fuels are of importance to the storage, distribution, and engine fuelling processes. The parameter ought to be included in the surveillance of the quality of diesel fuels, as indicated by laboratory tests and engine tests with respect to the evaluation of the antifoaming properties of diesel fuels, their correlations, and impact on selected performance parameters of vehicles.

Considering the above, the full analysis of diesel fuels, which is due before storage, ought to comprise tests of conformity with the requirements of applicable standard specifications, and be carried out in accordance with the principles described earlier for gasoline.

The short analyses of diesel fuels with a zero content of biocomponents ought to be performed at least once in 24 months, assessing the following parameters:

- color and appearance;
- density;
- fractional composition;
- flash point;
- coking number.

The control analyses of diesel fuels, to be performed at least once in 12 months, are intended to assess the following:

- color and appearance;
- density;
- flash point.

In carrying out the respective analyses, the rules discussed earlier in this chapter for gasoline apply.

CRITERIA FOR THE ASSESSMENT OF THE QUALITY OF HEATING OILS IN DISTRIBUTION, STORAGE, AND OPERATING CONDITIONS

Heating systems and devices are specific in terms of their designs and requirements with respect to heating oils. A definite majority of such systems, especially those with lower capacities, as well as their fuel storage tanks or reservoirs, are located underground and in lower floors of the buildings or facilities they provide heat for. Heating oils are kept in suitable storage tanks which are supplemented once or twice a year, therefore, they must not have any content of flammable or explosive or malodorous material. Flammability is limited by the flash point (not lower than 55°C). In this aspect, favorable materials include vegetable oils (flash points higher than 295°C) and spent oils.

Compared with light petroleum oils, products of thermal destruction of spent oils and vegetable oils are characterized by much lower flash points. Alcohols have even lower flash points, which definitely limits their usefulness as fuels or fuel components.

There are no commonly used methods for the assessment of liquid fuels, especially alternative fuels. The concentration of those substances or groups in heating oils which are accountable for their noxious smell is determined instrumentally (by chromatography) though, as a rule, such assessment is carried out organoleptically. Petroleum based heating oils usually have rather low volatilities and moderately noxious smell. A definitely undesired odor is observed in spent oils, products of their thermal destruction, and some hydrocarbon waste. The most desirable odor (very low vapor volatility) is shown by fresh vegetable oils and their methanolysis products but this may change after lengthy storage.

Because fuel storage tanks are located in unheated rooms indoors, or outdoors (which is obvious for fire safety reasons), liquid heating fuels are required to have a specific low-temperature profile. It defines their fluidity, therefore, the possibility to transport the materials to storage tanks and then deliver them to the burners. If a given liquid heating fuel has unfavorable low-temperature properties, it is necessary to use depressants or heating devices. Compared with petroleum oils, worse low-temperature properties are shown by vegetable oils and their esters. A definitely optimum low-temperature profile is that of alcohols.

Long-term storage of heating fuels, especially when combined with their low chemical stability, favors sedimentation of all kinds of deposit and water. Sludge at the tank bottom may cause corrosion and eventually plug the fuel system. This leads to problems with repairing and cleaning the heating systems. Therefore, liquid heating fuels should be kept clean and stable, although the issue is frequently very much underestimated.

Heating systems running on oils are typically equipped with automatic control systems, therefore, they are started periodically, depending on heat demand. Especially in the heating season, boiler operating conditions favor the emissions of soot or other products of incomplete combustion of the fuel components. Particular attention is required in the case of aromatic hydrocarbons and heterocyclic compounds, especially chloroderivatives because all of them are hazardous compounds. Such conditions, as well as too low furnace temperatures, may promote the formation of polyaromatic compounds and dioxins.

That is why heating oils must not have a content of substances which cause this kind of phenomena, even though this is not directly stated in standard specifications. Particular attention is required in the case of flue gases resulting from the combustion of liquid heating fuels based on spent oils, even though a specific heating system may be adapted to such fuels.

Also important in selecting potential components for heating oils is the factor of their local availability and pricing. The distribution of petroleum heating oils is typically the domain of private businesses which follow the Western solution of offering domestically made heating oils or, if purchased abroad, delivering them to onsite using motor transport. The distribution of refineries or major heating oil stations or storage facilities is highly varied over the territory of Poland. Under the circumstances, heating oils supplies are unduly inexpensive because they have to be transported over long distances. What is more, Poland having no domestic petroleum resources depends on imports and on any disturbances in the petroleum market that may occur. Therefore, satisfying the demand on heating oils is becoming a major economic issue and determines the further growth of the heating industry based on heating oils. In view of the above, any potential components of liquid heating fuels, as safe and affordable substitutes for heating oils, ought to be considered also in the aspect of rational economics.

Key Requirements for the Quality of Heating Oils

A heating fuel is expected, first of all, to provide inexpensive heat energy as a result of its combustion in an environmentally-friendly process. Conventional petroleum-based heating oils are mixtures of hydrocarbons having various chemical compositions (carbon-to-hydrocarbon ratios) and structures. Consequently, their combustion may have different kinetics and results. Controlling the quality of heating oils is understood as the optimum blending of their chemical composition. The presence in heating oils of any other ingredients, in addition to carbon and hydrogen, is an unnecessary burden, which affects their quality.

In design works, selection of equipment and fuels as well as in the correct operation of heating systems based on liquid fuels, the knowledge of physico-chemical parameters of heating oils is a must. Most of such parameters – though only for heating oils – are set out in standard specifications for the respective fuel brands and ought to be confirmed in commercial product quality certificates.

Some of the essential properties of heating oils, of which the knowledge is indispensable in their production process, in designing heating systems, as well as in their sale and operation are the following:

- **Heat of Combustion**: Combustion is a chemical process in which heat energy is generated. The reaction takes place in a gas phase, in the presence of oxygen, at a temperature above the fuel's flash point value. The levels of hydrogen, carbon, and sulfur in the fuel are taken into account in calculating the generated amounts of heat and flue gases. Experimental determination of the heat of combustion (upper and lower values) is carried out in a bomb calorimeter. In practice, the lower heat of combustion, usually referred to as its "calorific value", has more practical use. The difference between the upper and lower values of heat of combustion is the value of the heat of evaporation of the water contained in the fuel and formed during its combustion. The lower heat of combustion, for instance, for light heating oils, is in the range (41...42) MJ/kg.

- **Specific Heat**: Its value is necessary for calculating the heat demand required to heat a liquid heating fuel. The parameter is expressed in [kJ/kg K], and can be calculated from Crag's empirical formula:

$$C_{Dt} = \frac{I}{\sqrt{\rho}}(0,403 + 0,00081t)$$

(16)

where:

ρ – specific density of fuel, kg/m^3

t – temperature of fuel, K.

On average, specific heats for liquid heating fuels at temperatures in the range (273...473)K are (1.7...2.0) kJ/kg K.

- **Heat of Evaporation:** Heating oils are complex mixtures, comprising a variety of hydrocarbons having different boiling ranges, and it is not possible to state an exact value of heat of evaporation. Therefore, its value is determined using empirical formulae (such as Trouton's): it is in the range (209...230) kJ/kg for light heating oils, (189...209) kJ/kg for medium heating oils, and (147...189) kJ/kg for heavy heating oils.

- **Heat Conduction:** This value is required for designing heating systems. It depends on the chemical composition and phase composition of heating oils, as well as on temperature and pressure. For light heating oils, heat conduction is 0.116 W/mK.

- **Viscosity:** Viscosity is essential for the correct spraying of fuel, it determines the quality of the combustion process. The value of viscosity of a heating oil has an effect on its droplet size during spraying. Spraying is effected by means of dedicated sprayers (burner nozzles). Every nozzle will release a jet of optimum fine droplets of a heating oil for its specific viscosities only. Heating oils during spraying should, preferably, have viscosities in the range (3...30) mm²/sec. Some nozzle designs are capable of spraying liquids with viscosities up to 45 mm²/sec. A majority of heating oils have much higher viscosities at ambient temperatures, therefore, their pre-heating is required. All standard specifications concerning heating oils relate to the maximum permissible values of viscosity. They are preferably determined at 40, 50, 80, and 100°C. The range is very wide: typically from 20 to 190 mm²/sec at 50°C (standards PN, GOST, BDS, TGL).

Assuming that the correct viscosity of heating oil (in mm²/sec) for different types of spraying devices is as stated below:

- low-pressure air burner (12.5... 18.5)
- medium-pressure air burner (15.5...24.0)
- vapor burners (21.0...29.0)
- pressure burners (15.5...24.0)
- centrifugal burners (21.0...34.0)
- injectors in self-ignition engines (12.5...24.0),

then heavy heating oils (with high viscosities) need heating to high temperatures. For instance, a heating oil of which the viscosity is 100 mm²/sec at 50 °C requires heating to 102, 92 and 79°C to have viscosities of 15, 20 and 30 mm²/sec, respectively.

- **Density:** It is useful in establishing the category and origin of heating oils, their combustion and spraying efficiency. It is essential in settlements of delivery-acceptance operations. Density of heating oils is expressed in different units and at different temperatures: this complicates the comparison of fuels in this aspect. For instance, in the standards PN, CNS, GOST, TGL and BD, density is stated at 20^0C (and is in the range $(0.910...1.015)$ g/cm³); in DIN – at 15^0C $(0.860...1.20)$ g/cm³, in ASTM in 0API $(30...35$ range of values).

- **Flash Point:** It determines fire safety in the aspect of storage and use of heating oils and their heating limit. The minimum flash point of heating oils is stated in all standard specifications. Its values are in the range 65-85°C (PN, GOST, CNS, TGL, BS, ASTM, DIN). The lowest flash points of heating oils are stated in standards applicable in Western countries (ASTM: $(38...60°C)$; BS: 38, 56 and 66^0C; DIN: 55, 65 and 85^0C; JIS: 60 and 70^0C, depending on grade). Some manufacturers provide different values for different heating oil types: light, medium, or heavy oil.

- **Fire Point:** It is the lowest temperature at which product vapors continue burning for some time after being set on fire in open cup (Marcusson method). The lighter the oil, the larger the difference between the fire point and the flash point.

- **Self-Ignition Temperature:** The lowest temperature at which the vapors of a heated heating oil will ignite spontaneously after being contacted with air. Self-ignition temperature of light heating oils is expected to be in the range $(600...630)K$. The parameter is essential for the safety of boiler rooms where liquid fuels are used, especially in conditions of incomplete combustion, which is a potential cause of fire.

- **Freeze Point:** The parameter determines the transport and storage conditions for heating oils. It depends on the group composition of such oils and is frequently connected with the type of feedstock. Heating oils and hydrocarbon fractions obtained from high-wax diesel fuel as well as heavy heating oils have high freeze points.

The value for fuel fractions of high-wax diesel fuel is 10...15°C higher than that for low-wax diesel fuel. In relevant standards in the West, freeze point is stated only for light and medium heating oils. The value of the parameter is in the range from-8 to 0°C for light heating oils and to+40°C for heavy heating oils.

- **Fractional Composition:** It is very important for the combustion of light liquid heating fuels because it reduces the content of the fractions which evaporate to a specified temperature, breaks flame continuity and leads to generation of soot. As the boiling range of a liquid heating fuel increases, a higher level of aromatic hydrocarbons is observed (as well as that of sulfur compounds and resinous compounds with the tendency to generate soot and to undergo coking). For the course of the combustion process to be correct, heavy liquid heating fuels must have a content of flammable fractions which provide energy for disintegration of the large droplets of heavy fractions. Fractional composition is only specified for light heating oils. In practice, values are measured only for the percentage by volume of those fractions which distill to a specified temperature or for the boiling range for a specific percentage of a liquid heating fuel.

- **Sulfur Content:** The possibility of high sulfur content in liquid heating fuels is directly related to the origin of their components and their purification and blending technologies. In boiler fuelling systems based on liquid fuels, the presence of sulfur has no harmful effect, although its oxidation to form SO_2 leads to corrosion in flue discharge systems and to environmental pollution. Therefore, sulfur content in liquid heating fuels ought to be kept consistently low. Sulfur content in top-quality liquid heating fuels must be a max. 0.3% (m/m). For fuel blends based on crude oil distillation products with low and medium sulfur levels, the value is in the range (0.5...1.0)% (m/m). Heavy fuels based on medium-and high-sulfur fractions have a sulfur content in the range (5...7)% (m/m), which leads to sulfur compound emissions into the atmosphere – to a larger extent than in the combustion of high-sulfur coal.

- **Ash Content:** Liquid heating fuels, especially heavy ones, have a – usually low – content of dissolved, bound chemical elements (organometallic compounds), which form ash on burning. Some of its components (for instance, vanadium pentoxide) may lead to high-temperature corrosion of boiler parts. Ash particles, as they are carried into the atmosphere, tend to intensify solid emissions. Therefore, the parameter is limited in standard specifications to 0.1% (m/m).

- **Content of Foreign Solids:** Liquid heating fuels may have a content of solids originating from the production process or penetrating during distribution and storage. Sedimentation of the solids is a very slow process because the fuels are highly viscous substances. Therefore, it is necessary to remove the solids by filtering or centrifuging the fuels after heating but before delivering them to the burner, sprayer or injector. Since these are laborious and energy-consuming steps, it is more advisable to limit the content of mechanical impurities in liquid heating fuels. The maximum permissible values of that parameter are stated in all standards for heating oils. In liquid materials, such values should be different depending on its grade and application. The permissible content of solids is in the range (0.01...0.1)% (m/m) for light heating oils and (0.2...0.5)% (m/m) for other ones.

- **Water Content:** Water in liquid heating fuels may also appear in production processes, during distribution, and storage: either dissolved, or in the form of emulsion, or free water. Its presence in fuels leads to sludge formation processes and, consequently, to disturbances in the work of the boiler fuelling systems. The total content of free water (including emulsions) is determined by the distillation method.

- **Vanadium Content:** Vanadium content is determined because of the potential risk of high-temperature corrosion of furnace systems in boilers fuelled with liquid heating fuels of which the initial boiling range is above 300°C, originating from the blending of heavy waste hydrocarbon fractions. The lighter range of liquid heating fuels and mixtures of non-refinery raw materials have a negligibly low vanadium content so there is no need to assess the value in them.

- **Tendency To Form Carbon Deposit:** It characterizes fuel's ability to form solid residues after liquid-phase evaporation in the absence of air. The phenomenon potentially occurs if the stoichiometry of the combustion process is disturbed (incorrect spraying and evaporation of fuel, locally absence of oxygen). An increased tendency to form carbon deposits is observed when switching from lighter to heavier liquid hydrocarbons. The parameter is assessed by measuring the carbon residue.

For the full analysis of heating oils, the same rules as those provided in standard specifications for the other fuels may be applied. The short analysis ought to be performed at least once in 12 months to assess the following parameters:

- flash point;
- density;
- water content;
- viscosity;
- flowability or freeze point.

The control analysis ought to be performed at least once in 6 months, to assess the following parameters:

- flash point;
- density;
- water content.

The same rules of procedure as those set out for the other fuel types should be followed for the respective types of analysis.

THE QUALITY OF FUELS, AS ILLUSTRATED IN REQUIREMENTS OF COMBUSTION ENGINE MANUFACTURERS

Developments and modifications relating to the propulsion systems of the present-day engines depend, among other things, on technological progress in the area of suitable fuels for such propulsion systems. It is necessary to ensure compliance with applicable requirements connected with such processes as fuelling, evaporation, combustion, and dealing with exhaust gases, as well as those connected with environmental protection, transport, storage, and fuel distribution. Therefore, fuels must be in conformity with requirements concerning fuel combustion devices. The requirements are set out by design engineers and vehicle manufacturers, to guarantee sturdy and reliable engines, running on the suitable types of fuels.

The World-Wide Fuel Charter (WWFC) was established by major global vehicle manufacturers to assure the appropriate quality of fuels for spark-ignition and self-ignition engines and their conformity with combustion engine design requirements and environmental protection laws.

The most recent, Fifth Edition of WWFC, came into force in September 2013. It classifies fuels into five categories, both in the spark-ignition and self-ignition engine fuel categories.

Category 1 comprises requirements applicable to fuels used in markets with no or minimum requirements (US Tier 0, EURO 1) on the control of harmful emissions in exhaust gases; such fuels comply, first of all, with the essential requirements for engines and vehicles.

Category 2 comprises fuels being used in markets with more severe general and environmental-protection requirements, including those conformable with US Tier 1, EURO 2/II, EURO 3/III or equivalent requirements relating to toxic emissions in exhaust gases.

Category 3 comprises fuels for use in markets with severe general requirements and those relating to the control of toxic emissions in exhaust gases, conformable with US LEV, California LEV, ULEV, EURO 4/IV and P 2005 or equivalent, with regard to toxic emissions in exhaust gases.

Category 4 comprises fuels with more severe quality requirements, especially with regard to emissions, inclusive of advanced methods to reduce NOx and solids in exhaust gases. Category 4 fuels are in conformity with US Tier 2, US Tier 3 (pending), US 2007/2009 Heavy Duty On-Highway, US Non-Road Tier 4, California LEV II, EURO 4/IV, EURO 5/V, EURO 6/VI and JP 2009 requirements or equivalent standards; in addition to requirements relating to fuel consumption reduction.

The Fifth Edition of WWFC, introduced a new category of fuels, Category 5, which comprises those conformable with highly advanced requirements for emissions control and fuel efficiency, including US 2017 light duty fuel economy, US heavy duty fuel economy, California LEV III or equivalent standards, including those applicable to Category 4.

The WWFC requirements apply to a finished product, therefore, if a fuel is conformable with such requirements, no extra requirements or methods for intermediate or internal control are necessary or should be introduced.

Those WWFC provisions which are applicable to spark-ignition engines indicate the following essential tendencies in the development of such fuels, in connection with requirements applicable to engines ad vehicles:

- Reduction of sulfur level in fuels, because its combustion products have a toxic effect on the environment, and sulfur tends to poison the catalytic converter, thus limiting the possibility to reduce the amount of hydrocarbons NOx in the exhaust gases. In addition to that, sulfur affects the efficacy of elimination of NOx in prospective solutions based on combustion of lean blends (fuel economy).

- Elimination of the content of lead compounds;

- Elimination of additives which may promote ash formation;

- Elimination of the content of octane-enhancing, organometallic compounds such as methylcyclopentadienyl manganese tricarbonyl (MMT) or (iron-based) ferrocenes. Studies are in progress to establish the effect of manganese compounds on the efficiency of catalytic converters. It is assumed that manganese compounds are comparable with iron compounds in that they tend to form deposits on the surface of catalytic converters, thus increasing the content of toxic components in exhaust gases;

- Elimination of silicon and silicon compounds, which may be introduced into fuels by blending them with various spent solvents, causing failure of the oxygen sensors and catalytic converters;

- Limitation of the introduction into gasoline of oxygenates such as ethers or ethanol and higher alcohols and exclusion of methanol from fuels. The limitation results from the degraded driveability and inefficient reduction of NOx emissions in engines fuelled with lean blends. Moreover, tests of combustion of gasoline blended with 10% ethanol have shown that toxic emissions were 2% lower and carbon emissions were 10% lower, compared with a gasoline blended with 11% MTBE, although NOx emission was 14% higher, hydrocarbons 10% higher, and the ozone forming potential was 9% higher. This indicates that ethers are not permitted as an additional component, when blending gasoline with alcohols. Methanol is not permitted because of its potential to cause corrosion of metal components and degrade elastomer and other plastic components;

- Limitation of the use of olefins (unsaturated hydrocarbons), in spite of their octane-enhancing effect; this is justified by their potential to form resins and precipitate deposits and increase emissions of reactive hydrocarbons potentially leading to the formation of ozone and toxic compounds;

- Limitation of the content of aromatic hydrocarbons which favor the formation of deposits on engine components and lead to higher emissions of exhaust gases, including carbon dioxide;

- Limitation of the content of benzene because of its strongly carcinogenic effect;

- Formulation of a relationship between evaporation temperatures of 10%, 50%, 90% of fuel in normal distillation (T10, T50 and T90, respectively) and the percentage of oxygenates in the fuel (%OXY) to obtain Distillation Index (DI) as follows: DI=1.5 T10+3 T50+T90+11 %OXY. The ASTM standard also specifies the Driveability Index (DI) for gasoline containing ethanol according to the relationship: DI=1.5+3.0 T10 T50 T90+1.0+1.33 ° C (2.4 ° F) × Ethanol Volume%. DI values above the range 550...570 degrade driveability and hydrocarbon emissions are also higher. As seen, both of these indexes depend on the volatility and the content of oxygen derivatives.

As regards fuels for self-ignition engines, engine manufacturers indicate the following tendencies in the research works being made to improve the fuels quality:

- A minimum differential between the cetane number CN and calculated cetane index CCI is set, to avoid the excessive use of cetane-enhancers;

- A density range is set for each fuel category, as indicated by studies intended to find a correlation between the effect of lower densities and lower emissions of solids and NOx from high duty engines, with the accompanying lower engine power output and higher fuel consumption;

- A fuel viscosity range is established, as indicated by studies on the fuel injection process and fuel tank filling, which depends on that value, with the accompanying exponential decrease in viscosity due to the temperature increase;

- The permissible sulfur content is very much reduced in fuels for advanced engines because the presence of sulfur in fuels leads to higher solids emissions, stimulates low-temperature corrosion, and poisons NOx adsorbers, in addition to forming sulfur oxide during combustion;

- The content of polycyclic aromatic hydrocarbons (PAH) in fuels is reduced, even though their cetane numbers are low. This is due to their effect of increasing solids emissions (smoke level) and the presence of PAH in exhaust gases, while leading to an increased rough-engine operation tendency;

- Optimum values of distillation temperature for 90 or 95% (V/V) of fuel are established, as indicated by tests, because lower values cause lower emissions of NOx and higher emissions of hydrocarbons in the exhaust gases from high-duty engines, as well as a reduced emission of solids and higher emissions of NOx for light-duty engines;

- If low-temperature properties of self-ignition engine fuels are set from cloud points (CP) or from the Low Temperature Flow Test (LTFT) which is applicable in the USA and Canada, then the value of that parameter ought to be above the minimum expected ambient temperature. For the purposes of determination of Cold Flow Plugging Point (CFPP), then maximum value of CFPP ought to be equal to or below the lowest expected ambient temperature, except that the value should not be more than 10^0C above the set value of CFPP for the specific fuel category according to WWFC;

- The content of vegetable-derived esters, especially FAME (fatty acid methyl esters) in diesel fuels is limited because of their poor low-temperature properties, hygroscopic nature, higher tendency to form deposits, and aggressive effect on rubber seals and some other components in fuel systems. Oxidation stability of diesel fuels is established by Method 1 referred to in ISO 12205 and ASTM D 2274, although additional methods apply for fuels with more than 2 % (v/v) FAME, namely: Method 2a (modified Rancimat) according to EN 15751, Method 2b (Δ TAN) according to ASTM D 664, and modified ASTM D 2274 and Method 2c (PetroOxy) according to EN 16091;

- Injector fouling is controlled; the parameter depends on the amount and types of detergents being used as additives to improve fuel atomization, among other things;
- Lubricity of fuels is determined in order to ensure the appropriate operation of the injection pump while eliminating sulfur from the exhaust gases. Its value is established, at 60°C, using the High Frequency Reciprocating Rig (HFRR) testing device.

MILITARY TESTS APPLICABLE TO FUEL STORAGE IN THE NATO MEMBER STATES

The fuel quality assessment criteria that are applicable in the military technology of NATO member-states are based on long-standing experience. They are highly rigorous, therefore, they provide the most reliable criteria that guarantee the desirable quality of petroleum products in the contemporary military technology. The currently applicable standard is STANAG 3149 SILCEP (Edition 8) "Minimum Quality Surveillance of Petroleum Product" and was introduced in September 2002.

In STANAG 3149 SILCEP, all the applicable analytical tests were categorized as follows:

- Test A: An exact specification of all relevant parameters according to the standard. Applicable before product acceptance from supplier. Required for all storage tanks (with the exclusion of airfields) after filling them for a first time, for storage tanks after fuel replacement, and for storage tanks after cleaning.
- Test B-1: Applicable on completion of product reloading (transfer) using a non-segregating system such as multi-product tankers, or a pipeline system, or a common loading system.
- Test B-2: Applicable in fuel quality assessment after a pre-defined period of storage.
- Test B-3: Applicable before starting the transfer/flow of next product batches through the system in which they are not

separated, before combining product batches, and in tanks which hold mixtures from pipelines before repumping.

- Test C: Visual test, intended to make sure no visible changes have occurred in product. Applicable mainly for the routine assessment of fuels' quality (storage tanks, pipelines, tank trucks, railway tankers).

The scope of determination applicable to each test for the respective fuel types is shown in Tables 8, 9, and 10.

Table 8: Scope of determination in tests for gasoline

Characteristics	test			
	B-1	**B-2**	**B-3**	C
Appearance	X	X	X	X
Content of water and solids	X	X	X	X
Color	X	X	X	X
Density	X	X	X	X
Fractional composition	X	X	X	-
Vapor pressure	X	X	-	-
Corrosion on copper	-	X	X	-
Inherent resin content	-	X	X	-
Oxidation stability	-	X	-	-

Table 9: Scope of determination in tests for diesel fuels

Characteristics	test			
	B-1	**B-2**	**B-3**	C
Appearance	X	X	X	X
Color	X	X	X	X
Density	X	X	X	X
Fractional composition	X	X	-	-
Flash point	X	X	X	X
Coking value	X	X	-	-

Table 10: Scope of determination in tests for heating oils

Characteristics	test			
	B-1	B-2	B-3	C
Flash point	X	X	X	X
Density	X	X	X	X
Water content	X	X	X	X
Viscosity	X	X	-	-
Flowability, or freeze point	-	X	-	-

CONCLUSIONS

A Fuel Quality Assessment System is indispensable in the aspect of correct management of fuel storage, and its consistent application is required to prevent measurable losses. Improvements in the system ought to be made by eliminating redundant determinations of those qualitative parameters of fuels which do not change in storage conditions which are correct in the aspect of technology and performance. It is obvious that the parameters ought to be established by way of analyses, and be unambiguous. The currently applicable methodology of testing the quality of fuels during storage is based on measurements of a number of parameters which are indicated in the applicable standard specifications: this requires the development and use of a suitable procedure for sample collecting and their analysis in accredited laboratories. The frequency of sampling and the sample sizes were established based on previous experience in fuel storage. As shown in Chapter 1, fuel blends are subject to changes resulting from technological progress in the manufacturing of the respective components, specifically, in the area of technological processes based on the use of novel processing methods for handling heavy ends and residues in order to obtain fuel cuts or components. The use of new fuel components may change their oxidation kinetics, thereby changing the scope of quality surveillance which is indispensable in establishing the permissible duration of fuel storage. Therefore, in order to improve the cost-effectiveness and complexity of existing quality surveillance systems, storage facilities may be equipped with rapid on-line analysis kits enabling surveillance of the quality of (a number of parameters of) engine fuels, so that control analyses can be carried out; the findings would then be an early warning, indicating tendencies of changes in the quality of the fuels during storage but would also, in undisputable cases, suggest the necessity to determine certain parameters using more sophisticated, laboratory methodology, or to carry out a full analysis – this, altogether, could improve the cost-effectiveness of the Fuel Quality Assessment System.

Chances are that the course of oxidation of such mixtures will be changed entirely by the introduction of biofuels as additives to conventional fuels. Although so called "drop-in biofuels", which are compositions of hydrocarbons obtained by biomass or waste material processing in BtL (Biomass-to-Liquid) or WtL (Waste-to-Liquid) processes, have no significant effect on the kinetics of oxidation reactions, the course of oxidation of fuels which have a content of fatty acid alkyl esters (FAME, FAEE) as well as alcohols and ethers used as engine fuel components is changed entirely by such additives. Hence, considerations described in this monograph refer to an analysis of the findings of tests intended to develop and construct a system enabling the continuous monitoring of the engine fuel ageing process, including fuels with a content of biocomponents such as ester and other hydrocarbon derivatives. The project to implement the system is intended to enable on-line control of the operating usefulness of engine fuels during storage.

The above notwithstanding, it is advisable to develop and adopt applicable standard specifications for fuels which are intended for medium-or long-term storage.

REFERENCES

1.   K.Biernat „Prognoza rozwoju paliw" [A forecast for the development of fuels] „Studia Ecologiae et Bioethicae" No. 3/2005;

2.   K.Biernat, A.Kulczycki „Prognoza dla Polski" [A forecast for Poland] „Nowe Życie Gospodarcze" special supplement of 26 April 2006;

3.   K.Biernat, „Kierunki rozwoju układów zasilania i spalania oraz źródeł energii we współczesnych silnikach spalinowych" [Trends

in the development of fuelling and combustion systems and energy sources in contemporary internal combustion engines], „Przemysł Chemiczny" December 2006;

4. J.Merkisz „Ekologiczne aspekty stosowania silników spalinowych" [Environmental aspects of the use of combustion engines], Wyd. Politechniki Poznańskiej, Poznań 1994.

5. J.M. Maćkowski, The EU Directives on Fuels Quality: a look beyond year 2000", conference proceedings: Mat. II Międzynarodowej Konferencji pt: "Rozwój technologii paliw w świetle Dyrektyw Europejskich i Narodowych Uregulowań Normatywnych" [2nd International Conference: The development of fuel technology in the light of European Directives and national normative regulations] Warsaw 1999

6. J.Gronowicz „Ochrona środowiska w transporcie lądowym" [Environmental protection in land transport], ITE Radom 2004.

7. L.J.Sitnik „Ekopaliwa silnikowe" [Environmentally-friendly engine fuels]. Oficyna Wydawnicza Politechniki Wrocławskiej. Wrocław, 2004.

8. K.Biernat, A.Łuksa „Węglowodory i tłuszcze odpadowe jako ciecze opałowe" [Hydrocarbons and fat waste as liquid heating fuels], „Sozologia" No. 1/2003.

9. J.M Maćkowski, „ Paliwa samochodowe początku XXI wieku" [Automotive fuels of early 21st century], Wyd. Centrum Badawczo-Szkoleniowe Diagnostyki Pojazdów Samochodowych", Katowice 2002.

10. Gray C., Webster G.: A Study of Dimethyl Ether (DME) as an Alternative Fuel for Diesel Engine Applications. Advanced Engine Technology Ltd. Report TP 13788E; Canada, May 2001;

11. Gronowicz J. „Ochrona środowiska w transporcie lądowym" [Environmental protection in land transport], Instytut Technologii Eksploatacji, Poznań-Radom 2004

12. K. Biernat „Materiały Pędne i Smary" [Propellants and lubricants], WAT, Warsaw 1983

13. K. Biernat, Z. Karasz „Paliwa płynne i ich użytkowanie" [Liquid fuels and their use], SIMP, Warsaw 1997.

14. K.Baczewski, T.Kałdoński, „Paliwa do silników o zapłonie samoczynnym" [Fuels for self-ignition engines], WKiŁ, Warsaw 2004.

15. K.Baczewski, T.Kałdoński, „Paliwa do silników o zapłonie iskrowym" [Fuels for spark-ignition engines], WKiŁ, Warsaw 2005.

16. A.Podniało, „Paliwa, oleje i smary w ekologicznej eksploatacji" [Fuels, oils, and lubricants in their environmentally-friendly operation], WKiŁ, Warsaw 2002.

17. Worldwide Fuel Charter, Fifth Edition, September 2013;

18. STANAG 3149 SILCEP (Edition 8) „Minimum Quality Surveillance of Petroleum Product"

19. K.Biernat, A.Malinowski, J.Czarnocka: Chapter No18 in Monograph Editor: Zhen Fang, Title: „Biofuels-Feedstock, Production and Application", Editions by "InTech", December 2012: „An analysis of physico-chemical properties of the next generation biofuels and their correlation with the requirements of Diesel engines".

Chapter 7

Reverse Engineering of Turbocharger Compressor Designs Based on Non-Parametric CAD Data

Oliver Velde[1], Gero Kreuzfeld[1], and Ingolf Lehmann[2]

[1]CFturbo Software & Engineering GmbH, Dresden, Germany
[2]Kompressorenbau Bannewitz GmbH, Bannewitz, Germany

ABSTRACT

The improvement of turbocharger components—such as compressor—by means of virtual methods can be done most efficiently if those component's geometries are given in a parameterized format. It is shown here how the geometric description of turbocharger compressors that have a neutral format (ASCII or STEP/IGES) can be transformed in a parameterized description. This description is then used to perform parameter variations which are validated via simulation methods like CFD. The parameterization of the geometry is therefore a very important step within the workflow of the virtual design of turbocharger components as without it the investigation of different geometries are very time consuming and expensive.

INTRODUCTION

CAD-data of components that have been produced over a long period are very seldom in a parameterized format which can be used in an optimization process. On the other hand those components have been proven to be well designed and documented, and are a good starting point for further developments.

If the aerodynamic behavior of a turbocharger compressor is to be improved especially fluid touched surfaces are needed in parameterized format. Parameters would be classical turbo machinery parameters like main dimension, blade angles, span wise description of blade geometry a. s. o. The transformation of blank CAD-data into a parameterized form is often time consuming and inexact.

It will be shown how any neutral format data of a turbocharger compressor can be transformed into a parameterized model fast and accurate. This model is afterwards exported into a CFD-package where the meshing is accomplished followed by a CFD-analysis. Therefore once the model is available, virtual methods can be used intensely.

Some comparisons with experiments are shown in order to validate the re-design. Also an example of the consequence of a parameter change on the basis of the re-design is given.

RE-DESIGN

Very often data of old designs are available only as ASCII-format. With the turbo machinery design software CFturbo these data can be transformed into a parametric CAD-model [1]. Other neutral formats, such as STEP or IGES, are also possible as reference.

The procedure of the re-design needs some user interaction in order to get an exact representation of the reference design. For the calculation of information values that can be used for the valuation of the suitability, the following parameter should be given:

- Fluid properties: as ideal gas or as real gas with a compressibility factor.
- Best point: mass flow, speed, pressure ratio.
- Inlet conditions: pressure and temperature.

The main dimensions, that are hub and suction diameter as well as outlet width and diameter, have to be taken out of the reference geometry. Those values can be easily retrieved from any CAD-package but are very often known a priori.

Other important parameters should also be given, i.e. the size of the tip, direction of rotation, number of blades as well as splitter blades.

The meridional contour can be loaded directly if they are available as z-r-data, see Figure1. These data can easily extracted from CAD data using a meridional section plane.

In CFturbo the meridional contour is represented by Bézier-splines. These splines have to be congruent to the loaded x-y-data in order to achieve an exact copy of the reference geometry, see Figure 2. The congruence can be made manually or automatically (Figure 3).

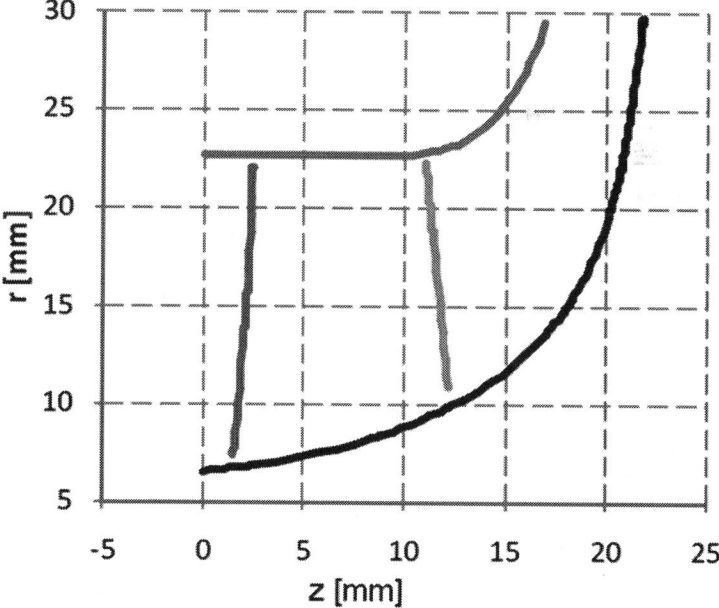

Figure 1: Meridional contour.

The shape of leading and trailing edge is straight by default but can be changed to freeform where again a Bézier-spline is used to represent any given form.

Blade angles: It is a strong advantage if the geometric variables can be extracted from the reference geometry in a way that they can be used directly in CFturbo. Such variables are e.g. radii, axial length, meridional and tangential co-ordinates as well as blade angles along mean lines, with whose help mean surfaces of the blades can be designed.

3D-data of any span are easily converted into variables (radius, meridional & tangential co-ordinate, blade angle) mentioned above by the following relations:

$$r = \sqrt{x^2 + y^2} \tag{1}$$

$$m = \int_{LE}^{TE} \frac{ds}{r} \tag{2}$$

$$t = \arctan\left(\frac{y}{x}\right) - t_{LE\,hub} \tag{3}$$

$$\beta = \arctan\left(\frac{dm}{dt}\right) \tag{4}$$

With these data all necessary variables are given for representation of the mean surfaces of the blades.

The mean surface is designed on the basis of mean lines lying in different spans. Mean lines can be given either as m-t-co-ordinates, or as β-t-co-ordinates. Again x-y-data of these pairs can be loaded into CFturbo followed by the match of the respective Bézier-spline description.

The blade profiling is the design step where a certain thickness is added to the mean surface on hub and shroud. Here almost any kind of profile can be realized as Bézier-splines with adjustable order as well as discrete profile data.

Some further finishing steps like leading edge rounding, trailing edge trimming, filleting a. s. o. may follow the re-design done so far.

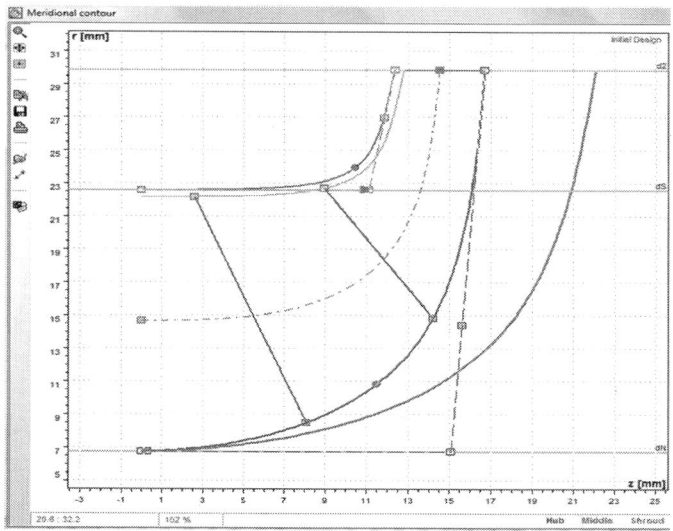

Figure 2: Initial meridional contour with loaded hub contour.

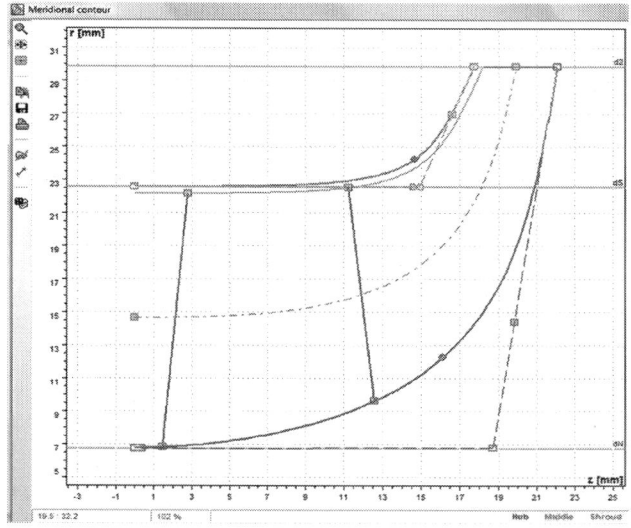

Figure 3: Matched meridional contour.

The procedure of the re-design of vaned and unvaned stators as well as of volutes is similar and is therefore not described here. The design steps are equally structured and allow both the calculation of parameters by CFturbo and their direct input.

PARAMETERIZED GEOMETRY

The geometric model is now a fully parameterized one because the crucial parameters are now fixed. Those parameters are given in Table 1.

Table 1: Design parameter

Name		Description
Main dimensions	d_N	Hub diameter
	d_S	Suction diameter
	d_1	Inlet diameter (leading edge)
	b_1	Inlet width (leading edge)
	d_2	Impeller diameter
	b_2	Impeller outlet width
	n_{Bl}	Number of blades
Meridional contour	$LE_{Hub,}$ LE_{Shroud}	Leading edge position on hub/shroud
	$LE_{HubSpl,}$ $LE_{ShroudSpl}$	Splitter leading edge position on hub/shroud
Blade properties	ProfCt	Number of blade profiles
	β_1, β_2	Blade angles at leading and trailing edge
	$\beta_{1,Spl}, \beta_{2,Spl}$	Splitter blade angles at leading and trailing edge

Mean lines	$d\boldsymbol{\varphi}_{Spl}$	Splitter trailing edge position (tangential) between neighboring main blades

EXAMPLE

The company Kompressorenbau Bannewitz (KBB) was founded in 1948 and has been active in the field of turbochargers (TC) since 1953. The main applications are medium-speed and high-speed 4 stroke diesel and gas engines [2-4].

Due to changed thermodynamic demands on the TC (two-stage-charging) imposed by the engine manufactures, pressure ratio per single compressor stage will decrease in the future. Stages of TC compressors produced 25 - 30 years ago, that were developed for similar pressure ratios, can serve as a good basis for further developments.

A certain compressor stage, comprehending of an impeller and a vaned stator, of the KBB-TC-family has been chosen. It was available as ASCII-data as well as CAD-model (see Figure 6, gold colored blades). The CAD-model was used as a criterion for the evaluation of the quality of the re-design geometry made by CFturbo. Both impeller and vaned stator show very good conformity.

From the available reference geometry the following data apart from the main dimensions and the best point could be extracted as x-y-data:

Meridional contour:

* z-r

Blade:

* β-m
* θ-m
* blade thickness
* x-y projection

Due to the fact that the blade design on the basis of the modification of β-m-data is more sensible than that on the basis of -m-data, it was tried to match reference data with re-design design data with the help of β-m.

In Figure 4 the **β**-distribution of hub and shroud are given. The match with the reference design is assessed by the difference of these distributions of reference design and re-design, see Figure 5.

The pure geometric assessment of the quality of the re-design has either been done by the comparison of the surfaces that represented the designs or by the comparisons of 2D-data that are available a priori or could be extracted from the original geometry. Since these data, especially the blade data like -m or -m, are very much dependent on the correct choice of a mean surface an entire consistency of all 2D-data is very difficult to gain.

COMPARISON WITH EXPERIMENT

For the reference design apart from geometric information measured data are also available. These have been used for the comparison with the redesigns simulation data (see Figures 7-8).

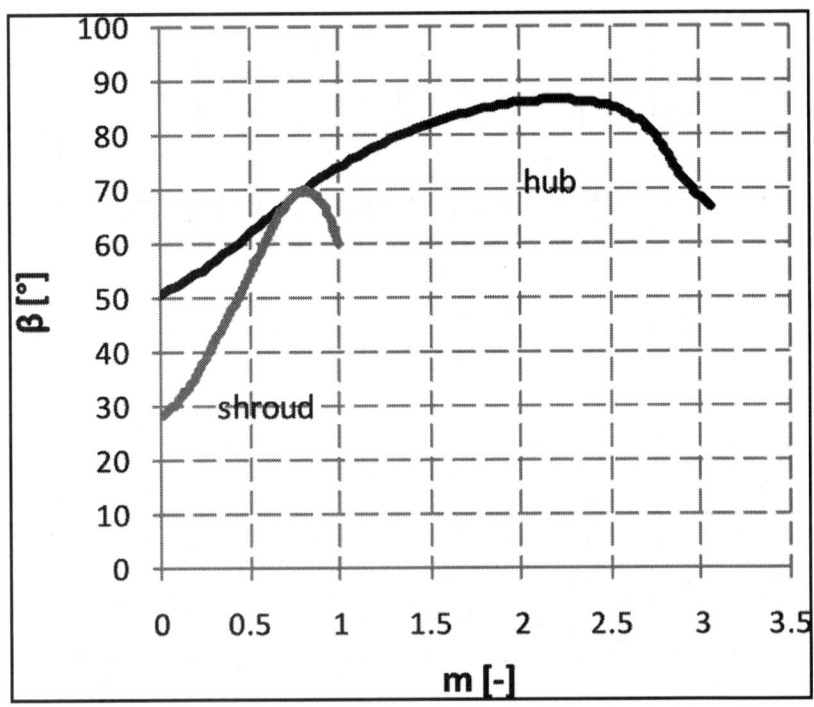

Figure 4: β-Distribution (re-design).

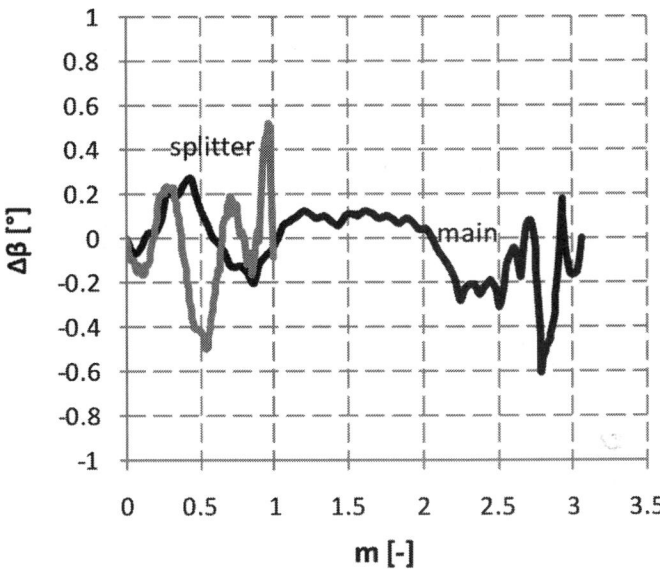

Figure 5: Differences between -distributions of reference design and re-design (only main blades).

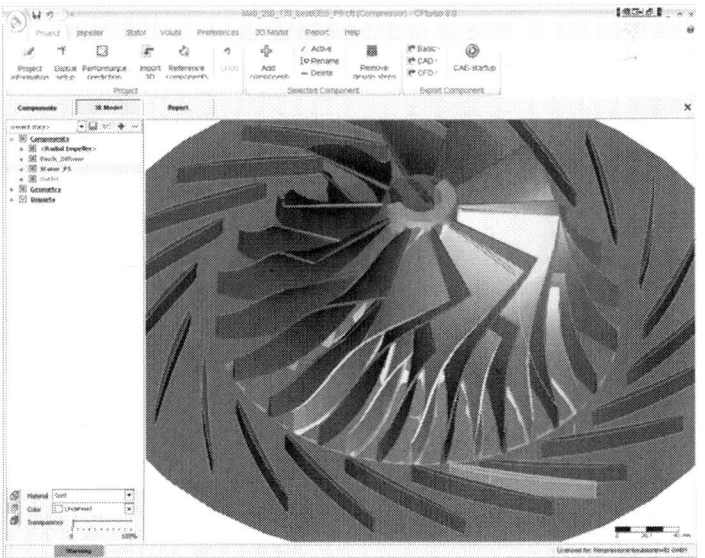

Figure 6: Reference and re-design in CFturbo (rotated).

One decisive difference between simulation and measurement setup of the compressor stage is that in the simulation a volute has not been considered. This explains especially at higher rotational speeds, the difference between measured and calculated efficiencies. The reason is the rising pressure loss in the volute at higher rotational speeds. The efficiency in the simulation is of course higher than those measured for the entire stage.

DESIGN VARIATION AND VALIDATION

For the validation of parameter variations made on the basis of the re-design 3D-CFD-simulations have been accomplished. To this end the CFD-interfaces integrated in CFturbo were used.

The general purpose CFD program NUMECA has been utilized to set up a simulation model of the segment of the stage comprehending of inlet pipe, impeller, vaned stator and radial outlet diffuser. This has been done for both geometries, reference and re-design respectively.

The parameters to be changed were the leading edge blade angles (see Table 2) because they were suggested by CFturbo differently. Obviously the blockage (blade thickness) is considered differently by CFturbo for the calculation of leading edge angles than that of the original design. This is shown by the big difference of the incidence at hub.

Almost all parameters to be chosen can be calculated on the basis of balance equations and empirical knowledge in CFturbo. Appropriate algorithms are implemented.

In Table 3 some of the integral results of the simulation are given. It is discernible that the influence of the leading edge blade angle is very small when they are changed in the range as chosen.

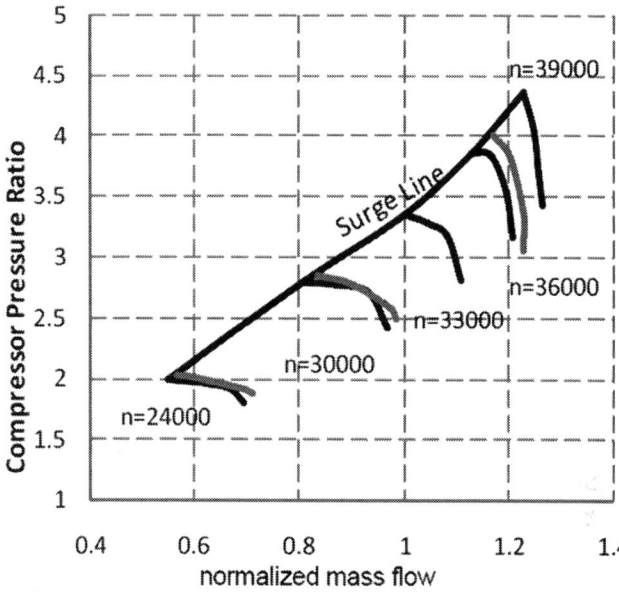

Figure 7: Comparison of measured (black) and simulated (grey) characteristics.

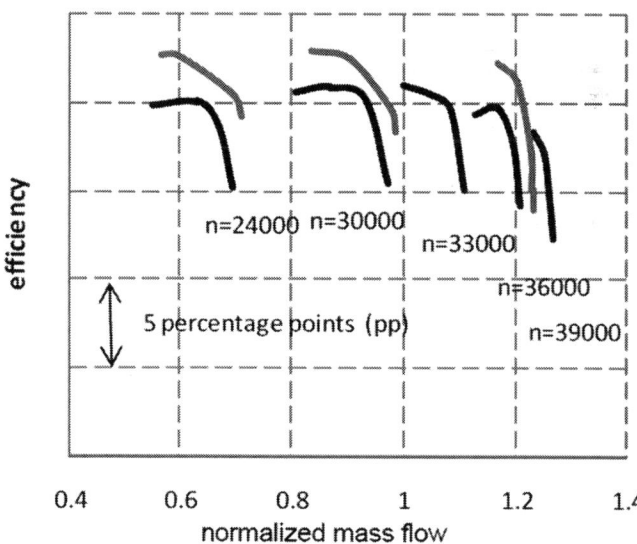

Figure 8: Comparison of measured (black) and simulated (grey) efficiencies.

Table 2: Original and changed blade angles

Name	Original parameter	Changed parameter
β_{B1} Main Hub [°]	54.7	65.2
β_{B2} Main Hub [°]	61.2	61.2
β_{B1} Main Shroud [°]	32.4	29.9
β_{B2} Main Shroud [°]	60.7	60.7
β_{B1} Splitter Hub [°]	79.1	79.1
Incidence Hub [°]	−11.2	0
Incidence Shroud [°]	2.5	0

Table 3: Simulation results, integral values

Name	Original geometry	Changed design
ϖ [-]	4.0545	4.0544
η_{stage} [-]	0.82374	0.82373
p_2 [Pa]	350220	350273

The analysis of the local flow fields shows almost identical results too. It is hardly discernible that differences exist at all (see for instance Figure 9).

Obviously the incidence changed in the given region has not a crucial influence on the aerodynamic. Certainly the thickness as well as the elliptic shape of the leading edge of the blades pay a contribution to this tolerance of the blade towards the change of the incidence.

Also closed looks at flow details do not reveal big differences (e.g. leading edge main blade at mid span, Figures 10 and 11 or 12 and 13 respectively).

The difference of the flow situation at the hub is much more noticeable because of the blade angle that has been determined with the blockage in design "changed" (see Table 2, local velocity plots at hub are not displayed). The development of the incidence from hub to shroud changes from negative to positive values. Very likely the incidence at mid span will therefore not be as big as at hub in the unchanged design, which might be another reason that the relative Mach Number distribution is very similar (see Figures 12-13).

The results of the simulations of both geometries yield the conclusion that the global aerodynamic performance of this type of compressor is hardly touched by a leading edge blade angle change.

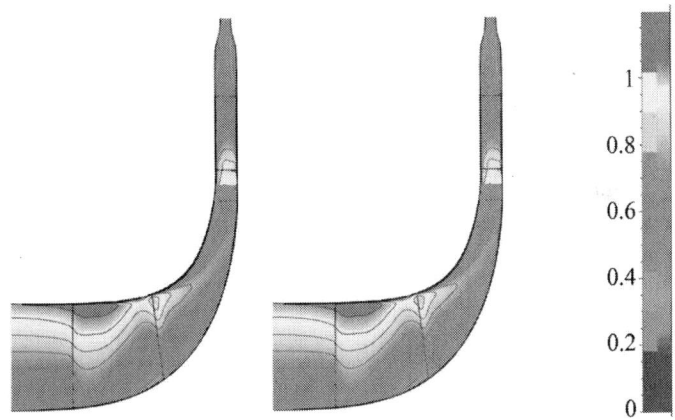

Figure 9: Relative Mach number in the meridional cut, left reference design, right re-design.

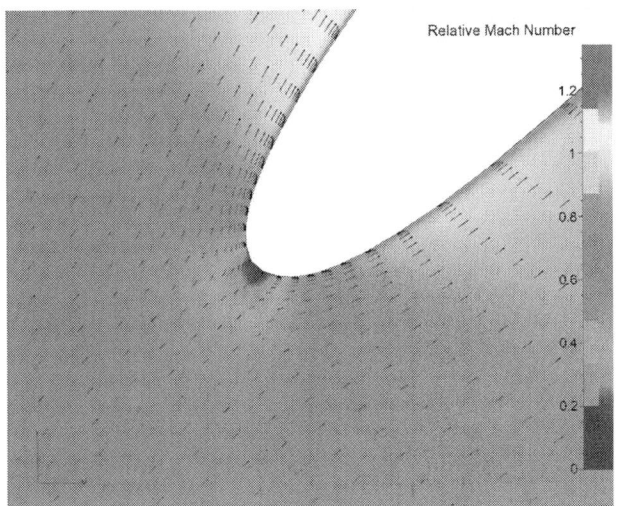

Figure 10: Mid span relative mach number original design.

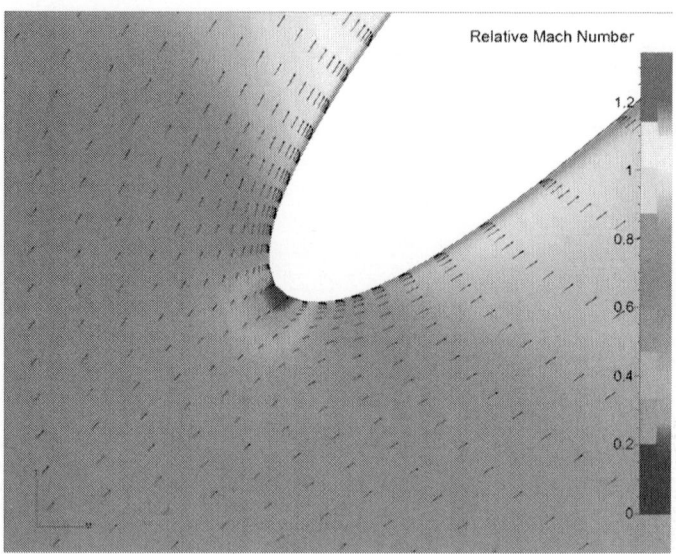

Figure 11: Mid span relative Mach number re-design.

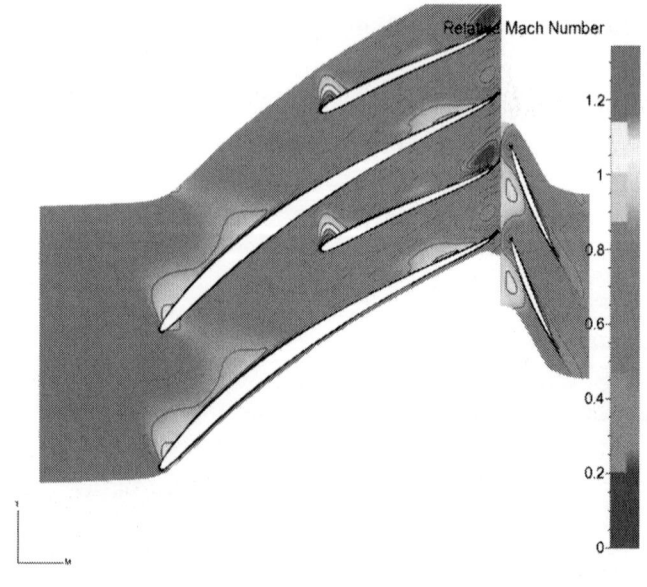

Figure 12: Mid span relative Mach number original design (blade-to-blade view).

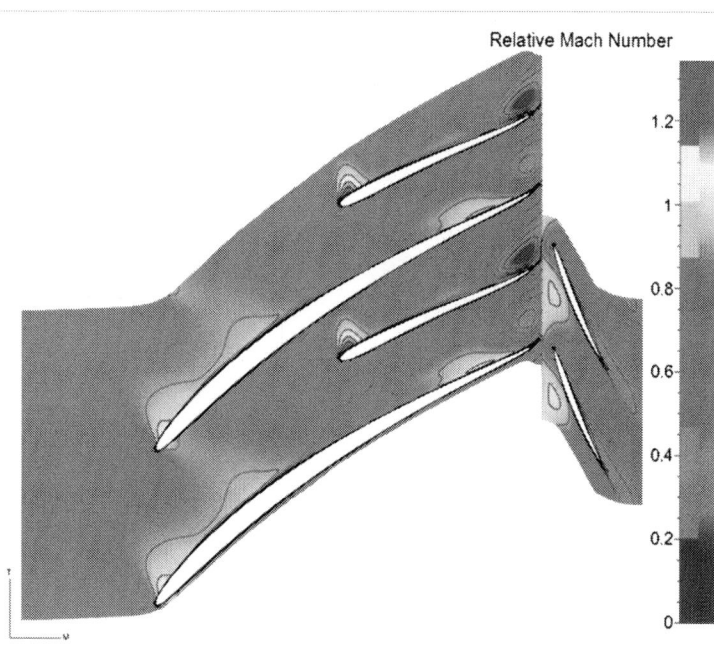

Figure 13: Mid span relative Mach number re-design (blade-to-blade view).

CONCLUSIONS

From the manufacturer's point of view the re-design process has been proven as exact and reliable. The parametric model can be used for derivative variants. Due to the utilization of CFturbo a broad data basis of most different impeller and stator sets can be created quickly. The parametric description of the geometries allows a very fast manipulation of geometries with respect to changing thermodynamic conditions like mass flow, pressure ratio or geometric restriction due to limited construction space.

The given example of a parameter change that was investigated numerically, shows the fast and efficient work flow that can be used once a fully parametric model is available.

REFERENCES

1.   G. Kreuzfeld and R. P. Müller, "An Advantageous Turbomachinery Design Method," Compressor Tech Two, August-September 2011, pp. 78-82.

2.   V. Tiede and U. Kramer, "New Compressor Design for Compact Turbocharger Range HPR from KBB," CIMAC Kongress, Hamburg, 7-10 Mai 2001.

3.   I. Lehmann, K. Buchmann and S. Käseberg, "HPR Range in Series Use—Ongoing Development of KBB Radial Turbine Type Turbochargers," CIMAC Kongress, Wien, Paper Nr. 102, 21-24 Mai 2007.

4.   I. Lehmann, K. Buchmann and R. Drozdowski, "ATLKomponentenentwicklung für Druckverhältnisse **über** 5," Presented at 12. Aufladetechnischen Konferenz Dresden, 27-28 September 2007.

Chapter 8

Engine Condition Monitoring and Diagnostics

Anastassios G. Stamatis[1]

[1]Mechanical Engineering Department, Polytechnic School, University of Thessaly, Volos, Greece

INTRODUCTION

Any engine exhibits the effects of wear and tear over time. Several mechanisms cause the degradation and potential failures of gas turbines such as dirt build-up, fouling, erosion, oxidation, corrosion, foreign object damage, worn bearings, worn seals, excessive blade tip clearances, burned or warped turbine vanes or blades, partially or wholly missing blades or vanes, plugged fuel nozzles, cracked and warped combustors, or a cracked rotor disc or blade.

Fouling is caused by liquid or solid particles accumulated to airfoils and annulus surfaces. Deposits consist of varying amounts of moisture, oil, soot, water-soluble constituents, insoluble dirt, and corrosion products of the compressor blades material whish are held together by moisture and oil. The result is a build-up of material that causes increased surface roughness and to some degree changes the shape of the airfoil. Hot corrosion is the loss or deterioration of material from flow path components caused by chemical reactions between the component and certain contaminants, such as salts (for example sodium and potassium), mineral acids or reactive gases (such as hydrogen sulfide or sulfur oxides). Corrosion is caused by noxious fumes or ash-forming substances present in the fuel such as aluminum, calcium, iron, nickel, potassium, sodium, silicon, magnesium. Corrosion increases surface roughness and causes pitting. Erosion is the abrasive removal of material from the flow path by hard or incompressible particles impinging on flow surfaces. Damage may also be caused by foreign objects striking the flow path components (Figure. 1a). Foreign Object Damage (FOD) is defined as material (nuts, bolts, ice, birds, etc.) ingested into the engine from outside the engine envelope. Domestic Object Damage (DOD) is defined as objects from any other part of the engine itself.

Different causes and mechanisms of performance deterioration of jet engines are reviewed in [1]. Degradation in both land and aero gas turbines is also reviewed by Kurz and Brun [2], who pointed out differences in mechanisms for the two types. Industrial gas turbine deterioration has been discussed by Diakunchak [3].

(a)   (b)

Figure 1: (a) FOD effects, (b) Turbine nozzles with deposits.

Three major effects determine the performance deterioration of the gas turbine compressor due to fouling: Increased tip clearances, changes in airfoil geometry, and changes in airfoil surface quality. In compressors, erosion increases tip clearance, shortens blade chords, increases pressure surface roughness, blunts the leading edge, and sharpens the trailing edge. Turbine blade oxidation, corrosion and erosion are normally longtime processes with material losses occurring slowly over a period of time. However, damage resulting from impact by a foreign object is usually sudden. Impact damage to the turbine blades and vanes will result in parameter changes similar to severe erosion or corrosion. Corrosion, erosion, oxidation or impact damage increases the area size of the turbine nozzle. When crude oil is burned in the GT the hot end is subjected to additional harmful deposits, including salt deposits originating in the inlet or from fuel additives. As hot combustion products pass through the first stage nozzle, they experience a drop in static temperature and some ashes may be deposited on the nozzle blades decreasing the nozzle area (Figure 1b). The combustion system is not likely to be the direct cause for performance deterioration. The combustion efficiency will usually not decrease, except for severe cases of combustor distress. However, plugged nozzles and/or combustor and transition piece failures will always result in distorted exhaust gas temperature patterns. This is a result of the swirl effect through the turbine from the combustor to the exhaust gas temperature-measuring plane. Distortion in the temperature pattern or temperature profile not only affects combustor performance but can have a far reaching impact as local temperature peaks can damage the turbine section.

All the above causes and effects may be considered as faults. Generally speaking, fault is a condition of a machine linked to a change of the form of its parts and of its way of operation, from what the machine was originally designed for and was achieved during its initial operation. In this respect a fault manifests itself by a change of geometrical characteristics or/and integrity of the material of parts of an engine. Change in geometry is inevitably linked to common experience faults, as for example when a part is broken, or deformed. Typical integrity fault is the occurrence of cracks inside the material, which are not associated to any geometrical change but can nevertheless result into catastrophic consequences. Some of the faults will become evident as vibration increases or by a change in lubrication oil temperature.

However, some serious faults can be detected only through gas path analysis. The gas path, in its simplest form, consists of the compressors, combustor, and turbines.

Diagnosis of a mechanical condition is the ability to infer about the condition of parts of the engine, without dismantling the engine or getting direct access to these parts, but only from observations of information coming to the engine exterior. The field of engineering science covering the techniques for achieving a diagnosis is called diagnostics. The aim of diagnostics is to detect the presence and identify the kind of faults appearing in a engine. Diagnostics does not require that the engine is either stopped or disassembled. Information is gathered while the engine is in operation. This is vital for engines in the process industry or energy production, as they must run without interruption for long time intervals. Detection of an incipient failure in a jet engine leads to taking action necessary to prevent a catastrophic failure which might follow.

In order to establish the possibility of diagnosing engine condition a correspondence of this condition to the values of the measured quantities should be known. In general terms, this correspondence is intrinsically established through the physical laws governing the operation of the machine. The behavior of any relevant physical quantity is linked through these laws to the detailed geometry of the machine and the kind of phenomena taking place in it. If we consider a machine using a fluid as a working medium, the variation of the flow quantities at one particular location in the machine is determined, via the laws of fluid mechanics, from the geometry of the solid boundaries and the physical properties of the fluid. A change in geometry will then reflect on the values of the flow quantities and could be calculated by application of the relevant physical laws. If suitable quantities are measured, they reflect changes in geometry or material and can therefore be used to indicate the presence of a fault. It is obvious that according to the change occurring in an operating machine, different quantities will be influenced. For example, the operation of rotating components is always linked to the exertion of periodic forces, with a frequency which is usually a multiple of the frequency of rotation. In this respect, the quantities characterizing a vibration are suitable for diagnostic purposes. On the other hand, severe corrosion, as it changes turbine airfoil geometry, is detectable through gas path analysis.

Many techniques for inferring engine status or change in engine condition have been proposed and/or applied to various engine configurations with varying success. Some of them (e.g. Vibration monitoring, Trending Analysis) are parts of computer-controlled data-acquisition systems that permit the on-line acquisition and reduction of a very large amount of performance information. While fault detection or general deterioration could be based on immediate observation of reduced measurable quantities, such observation is not, generally, adequate. It should also be noted that a change in any measured parameter does not necessarily indicate a particular independent parameter fault. For example, a change in compressor discharge pressure (CDP) does not necessarily indicate a dirty compressor. The change could also be due to a combined compressor and turbine fault or to a turbine fault alone. In order to have access to the variables, which possess diagnostic information (such as component efficiencies) modeling of an engine is essential. Thermodynamic (Gas Path) analysis methods employ engine models to process measurement data, in order to diagnose changes in component performance which may be linked to degradation, aging, or incipient failure.

GAS PATH ANALYSIS

An engine may be viewed as a system, whose operating point is defined by means of a set of variables, denoted as **u**. The operation of each component follows predictable thermodynamic laws. Therefore, each component will behave in a predictable manner when operating under a given set of conditions. The health condition of its components is assumed to be represented through the values of a set of appropriate "health" parameters such as efficiencies and flow capacities, contained in a vector **f**. The system is observed through measured variables, such as speeds, pressures, temperatures, contained in a vector **y**. When the engine operates at a certain operating point measured quantities are produced for given values of health parameters. The operating engine establishes a relationship between these parameters, which can be expressed though a functional relation:

$$y = \mathbf{F}(\mathbf{u}, \mathbf{f})$$

(1)

A computer model materializing this relation can reproduce the values of any thermodynamic quantity measured along the engine gas path. It is interesting to note that by assigning appropriate values to the components of vector **f**, the effect of engine component faults or deterioration on measured quantities can be reproduced.

The problem of diagnostics (Figure 2) is to seek a solution to the inverse problem, namely to determine the values of the estimated health parameters f̂ from a given set of measurements using a diagnostic method (*DM*). Particular faults can then be detected if deviations of health parameters from the reference state are observed.

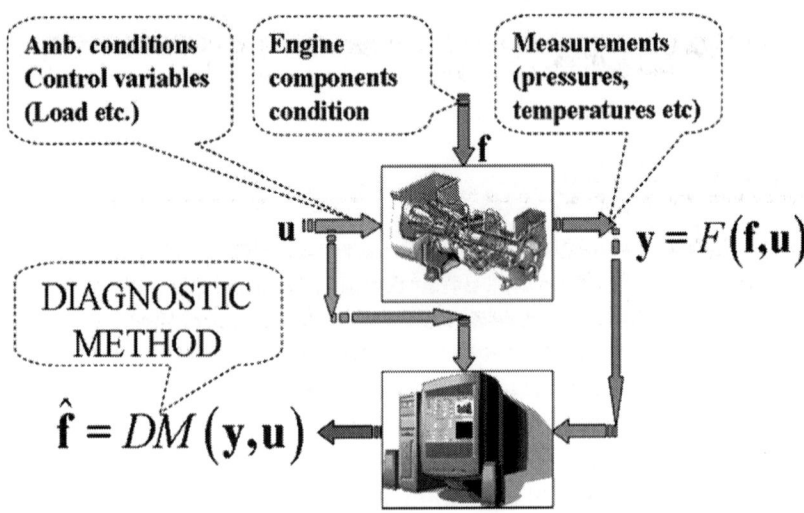

Figure 2: Gas Path Analysis diagnostics formulation.

Many variants of Gas Path Analysis based diagnosis with different features and complexity have been developed and reported in the open literature. Extensive reviews of existing methods provided by Li [4], and Marinai et al. [5].

Generally speaking any GPA method at least consists of the following elements:

- Measured data
- A data processing model relating measured data with health parameters
- A diagnostic decision making procedure.

The data used can be taken in steady state or transient operation. The model could be a physical one representing the aerothermodynamic processes taking place in the engine components and the mechanical coupling between them or a black box mathematical model relating data with health parameters. The diagnostic decision making procedure may be a conventional pattern recognition technique applied to health parameter space or an artificial intelligence based expert system.

Accordingly the proposed methods are classified on the basis of the kind of the comprising elements as: Steady state or Transient, Physical or Mathematical, Conventional or Artificial intelligence method.

PHYSICAL MODELS BASED GPA

Linear Methods

In linear gas path analysis, the health parameters are represented as the unknown "deltas" of component performance parameters (typically efficiency and mass flow capacity). They are related to known measurement "deltas" through relations produced by linearization of the general nonlinear thermodynamic relations, assuming small deviations. [6].The classical linear approach is formulated as follows: For a given operating point u the measurement values depend only on the health condition of engine components. After linearization and taking into account measurement uncertainty (by adding a noise vector **v** with zero mean and known covariance R), the typical GPA equations take the form:

$$\Delta y = C \cdot \Delta f + v$$

(2)

where Δ is called delta and represents percentage deviation from a reference value(when the engine is in intact condition) and C the well-known influence coefficient matrix. Estimation of health parameters is obtained from the relations

$$\Delta \hat{\mathbf{f}} = \mathbf{S}^{-1} \cdot \mathbf{C}^T \cdot \mathbf{R}^{-1} \cdot \Delta \mathbf{y}$$

(3)

$$\mathbf{S} = \mathbf{M}^{-1} + \mathbf{C}^T \cdot \mathbf{R}^{-1} \cdot \mathbf{C}$$

(4)

where M represents known statistics for the deviation of health parameters.

Although the formulation for classical GPA has proven to be successful for practical purposes and existing commercial systems ([7, 8]) are based on it, identifiability problems exist due to limited instrumentation. Sufficient engine health assessment requires at least the estimation of the parameters associated with the main engine components. Considering an existing engine, a typical situation is characterized by the fact that the number of available sensors is smaller than the number of parameters to be calculated. Accordingly, all the initially implemented methods were compelled to adopt various assumptions. Most of the methods use a priori information about the statistics of the calculated parameters introducing thus bias in the estimation. In that case, inversion of matrix S is only possible when it is dominated by M. The main drawback is the effect discussed by Doel [9]. The algorithm tends to "smear" the fault over many components.

Multi Operating Point GPA

GPA Multi Operating Point Analysis (MOPA) methods have been developed trying to exploit information provided by the existing sensors when different operating points are considered. The origin for the multi operating point analysis (MOPA) methods was the Discrete Operating point GPA, introduced in [10].The method, based on information given by existing sensors when different operating points are considered, improved significantly the diagnostic effectiveness. The implementation of the method was an extension of the classical linear gas path analysis. MOPA methods though do not use a priori statistics for the parameters rely on the questionable assumption of non-varying health parameters. Other research groups applied the same principle for the nonlinear case, [11-13].

The linear implementation for the MOPA approach using NOP operating points is given by Eqs. (5)-(9).

$$\Delta \mathbf{y}_k = \mathbf{C}_k \cdot \Delta \mathbf{f} \quad k = 1, \mathrm{NOP} \tag{5}$$

$$\mathbf{C}_k = [c_{ij,k}] \tag{6}$$

$$c_{ij,k} = (\partial \Delta y_i / \partial \Delta f_j)_k \tag{7}$$

$$\Delta \hat{\mathbf{f}} = \mathbf{P}^{-1} \cdot \sum_{k=1}^{\mathrm{NOP}} (\mathbf{C}_k^T \cdot \mathbf{R}_k^{-1} \cdot \Delta \mathbf{y}_k) \tag{8}$$

$$\mathbf{P} = \sum_{k=1}^{\mathrm{NOP}} (\mathbf{C}_k^T \cdot \mathbf{R}_k^{-1} \cdot \mathbf{C}_k) \tag{9}$$

The so called information matrix P is crucial in the sense that its condition determines the diagnostic effectiveness. The condition of the matrix is represented by its condition number. Investigations concerning effects of both the number of operating points used and the 'distance' of the operating points on information matrices have been reported ([14]-[15]). Additional details on assessing identifiability in multipoint gas turbine estimation problems are given in [15]. Although all the works implementing the multipoint approach agree that the idea more or less improves the diagnostic effectiveness, there are also results (see [16]), indicating that the theoretically attainable multi-point improvements are difficult to realize in practical engine applications.

In order to understand the reasons for potential problems concerning diagnosis using a multipoint approach it is necessary to examine the underlying assumptions of the method. The main assumption of the method is that the 'deltas' concerning the health parameters remain constant with regard to change in operating conditions. This assumption is obviously true for some parameters (for example the parameter expressing the effective turbine area or the area of non-variable nozzle jet engine), but there are indications that for other parameters this is a week assumption. Several works ([3], [17]), have provided evidence that when deterioration is present, the deviations of parameters such as flow compressor capacity and efficiency change with the operating point. In fact different working-point means different aerodynamic conditions and, in this sense, efficiencies and flow capacities deltas can significantly vary with the operating condition. The resulting diagnosis risk is not only to imprecisely calculate the engine new state after some deterioration but even more to indicate as responsible for the fault the wrong component(s).

Recently a new variant of GPA method named Artificial Multi Operating Point Analysis (AMOPA) has been proposed [18]. The new method uses existing sensor information produced when artificial operating points are used close to an initial operating point by using different parameters for each operating point definition. Therefore the assumption that the 'deltas' of the health parameters remain constant is reasonable. The method proved to be capable of both isolating and identifying the fault in individual components.

Nonlinear Methods

In nonlinear methods, the full thermodynamic equations are treated directly without simplification. An example of such a method, the method of adaptive modeling introduced by Stamatis et al. [19], uses component maps "modification factors" as health parameters and solves for them through an optimization procedure applied to a function based on differences of the predicted and measured values. Variants of the nonlinear GPA have been proposed (see [20-22]), the main differences being the objective function formulation as well as the method used for the optimization. The more general objective function (OF) to be minimized was proposed in ref. [23]:

$$OF = \sum_{i=1}^{m} \left[\frac{y_i^{calc}(\mathbf{f}) - y_i}{y_i \sigma_{Y_i}} \right]^2 + C_A \cdot \sum_{j=1}^{n} \left| \frac{f_j - f_j^r}{f_j^r \sigma_{f_j}} \right| + C_S \cdot \sum_{j=1}^{n} \left[\frac{f_j - f_j^r}{f_j^r \sigma_{f_j}} \right]^2 \tag{10}$$

where n and m the dimensionalities of $\mathbf{f}$ and $\mathbf{y}$ correspondingly. The first term express the fact that the health parameters under estimation $\mathbf{f}$ must be such that the values of measured quantities $\mathbf{y}$ are reproduced as accurately as possible. The second and third terms ensure that the values of health parameters cannot be significant different from their reference, a fact resulting from experience. It is the addition of these terms that allows the derivation of a solution for $\mathbf{f}$, even when a smaller number of measurements is available. All *deltas* are weighted by the inverse of the standard deviation of the corresponding quantity. Weight factors *CA, CS* are also included, for the possibility to change the relative importance of the two groups of terms. The reference values $\mathbf{f}^r$ of the health parameters can be chosen to represent a 'best' guess of the values to be determined. From studies in estimation theory, it has been found that it is useful to include in the objective function a term of sum of absolute values, since this term may improve the numerical behavior of the estimation procedure by increasing its robustness (see [24]).

The way of determining the vector $\mathbf{f}$ for minimization of this function can take advantage of the physical characteristics of the problem to be solved. For example the fact that deviation of component efficiencies should not be positive could be formulated as a constraint in the optimization. In the case of slow deterioration tracking, the reference values can be chosen to vary slowly with time while a filtering procedure can be applied, taking advantage of the regular variation of component deviations, as described in [25]. For the case of individual component faults the fault usually affects one or two neighboring components.

All these features should be taken into account when formulating the diagnostic algorithm. The solution is obtained with the interaction of a non-linear engine performance model and an optimisation algorithm, as shown in figure 3.

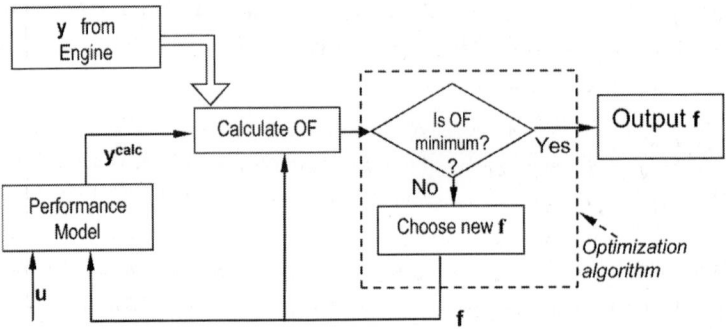

Figure 3: Schematic representation of nonlinear diagnostic procedure.

The methodology for diagnosing single component faults using the above procedure is based on the following reasoning. Since measurement data are noisy, the estimations based on a single data set differ from the actual values due to noise propagation. They can be improved when more than one measurement data sets are available. In such a case a solution is obtained for each individual data set. A series of values for each health parameter f_j becomes thus available. The mean value and standard deviation of the percentage change from reference $\Delta f j$ are then calculated. A criterion then is proposed for isolating the parameters of the components that are faulty, with the aid of a parameter, which we call diagnostic index. We define as diagnostic index the ratio of the absolute mean value to the standard deviation for each estimated health parameter.

$$DI_j = \frac{|\Delta f_j|}{\sigma_{f_j}}$$

(11)

Health parameters exhibiting small deviations from reference state or parameters with large standard deviations (large uncertainty on derived estimations) will have small values for diagnostic index. On the other hand, health parameters with large mean value or small standard deviation (small uncertainty on derived estimations) will present large values for diagnostic index. It is thus expected that the health parameters, which deviate due to fault occurrence will be those with the largest value of the diagnostic index. Thus we identify as faulty the component containing the parameter with the largest diagnostic index. This stage is called fault localization.

After the detection of a faulty component, a more accurate estimation of fault magnitude can be performed. The optimization problem is solved again by keeping as unknowns only the health parameters of the component found faulty. C_A, C_S are zeroed, to avoid biases imposed by the corresponding terms. (Note that with much fewer unknowns a unique solution can be derived by minimizing differences only from measurements namely the first term of the objective function eq (10).) After performing a series of estimations with this formulation from the available data sets, the average values of the obtained parameters are kept as the estimations for the fault magnitude.

Nonlinear GPA methods have proved accurate and robust provided that appropriate measured variables and estimated health parameters have been selected. This is not a trivial problem as explained in the following.

Sensors and Health Parameters Selection

When application of a GPA technique is envisaged on an engine, the existence of certain restrictions is recognized. Considering an existing engine, there is always a given set of available measurements. Addition of instrumentation can be difficult or even impossible. It is therefore important to have the possibility to adopt a convenient formulation of a method, so that an optimal use of the existing measurements is achieved. On the other hand, when a new engine is designed, or when an intervention to instrument an engine is performed, it is desirable to define an optimum combination of sensors to be installed.

The problems that may be faced in such a situation can be summarized as follows: When a given set of measured quantities is provided, what is the optimum set of health parameters? The particular problem is to define the best possible parameters for a given measurement possibility. This is a problem faced usually by the engine user, who has very few or no possibilities of intervening and adding measurements on the engine. When the decision for instrumenting an engine has to be taken, both the manufacturer and the user are faced with the inverse problem: (a) The user wants to know the optimum set of measuring instruments to be added in order to provide enough information for a required level of resolution. (b) The manufacturer wants to decide which instruments will accompany the engine, in order to ensure a good capability of in-service monitoring.

A systematic study for methods of choice of measurements and parameters in a way optimal as to diagnostic effectiveness was first presented by Stamatis et al. [26]. They introduced criteria for optimal measurement or health parameter selection. We present here the proposed method for measurement selection. Let $\mathbf{f}^{(r)}$ be the baseline diagnostic vector corresponding to a healthy engine (typically $\mathbf{f}^{(r)} = \mathbf{I}$), and $\mathbf{f}^{(j)}$ the diagnostic vector resulting when the jth element of $/$ deviates from the baseline value (f(r)) by a percentage amount h_j.

$$\mathbf{f}^{(j)} = \mathbf{f}^{(r)} \cdot + h_j \cdot \mathbf{e}^{(j)} \quad j = 1, ..., m \tag{12}$$

hj is a small constant (0.001 <hj<0.01). Then, from Eq. (3) we have

$$\mathbf{Y}^{(r)} = \mathbf{F}\left(\mathbf{f}^{(r)}\right) \tag{13}$$

$$\mathbf{Y}^{(j)} = \mathbf{F}\left(\mathbf{f}^{(j)}\right) = \mathbf{F}\left(\mathbf{f}^{(r)} + h_j \cdot \mathbf{e}^{(j)}\right) \tag{14}$$

The sensitivity of each dependent parameter on each individual health index is evaluated as

$$\Delta Y_k^{(j)} = \left(Y_k^{(j)} - Y_k^{(r)}\right) / Y_k^{(r)} \quad k = 1, ..., n \tag{15}$$

We also define an overall sensitivity measure for each parameter with the norm

$$SY_k = \left[\frac{1}{m} \cdot \sum_{j=1}^{m} \left(\Delta Y_k^{(j)} \right)^2 \right]^{1/2} \qquad k = 1, ..., n$$

(16)

So, the problem of selecting the appropriate measurements is expressed mathematically as follows: For a given set of health condition parameters, we must select as measured parameters these parameters giving on the norm of Eq. (16) the m greater values.

In later years more works have appeared, approaching the problem from different points of view, [14,27].

ARTIFICIAL INTELLIGENCE GPA METHODS

Neural Networks

An Artificial Neural Network (ANN) is an information processing paradigm that is inspired by the way biological nervous systems, such as the brain, process information. The key element of this paradigm is the novel structure of the information processing system. It is composed of a large number of highly interconnected processing elements (neurons) working together to solve specific problems. ANNs, like people, learn by example. Learning in biological systems involves adjustments to the synaptic connections that exist between the neurons. This is true of ANNs as well. Two tasks which can be performed by neural nets, which are relevant to the procedure of monitoring and diagnostics of a gas turbine, are: modeling the performance of a gas turbine and detection and classification of faults.

A typical use of a model is to produce reference values for quantities which are monitored. It can be also used for other purposes such as generation of influence coefficient matrices, and sensitivity analyses. ANNs are known to be able to model non-linear systems and therefore can be used for gas turbine performance modeling. A first advantage offered by modeling engine performance through ANN is the much shorter computational time required, once the net is trained and verified, in comparison to any full scale aerothermodynamic model. The latter involves the solution of a set of non-linear equations, which is achieved through iterative schemes, resulting in a number of arithmetic operations significantly larger than those performed by an ANN. A further advantage is related to the possibility of adapting to a particular engine, if data is available. A well-known fact is that for a model to be accurately representing the operation of an engine, it has to be adapted to the particular engine (as discussed, for example, in [19]). A model using ANN provides inherently this possibility, through the way it is being set up. The existence of a learning phase, (called "training" in the ANN terminology) allows the adaptation to a particular engine, if enough data is available.

The second area of possible application, detection and identification of faults, comes from one of the most powerful capabilities of ANN, namely the capability of identifying and classifying patterns. Any method of fault detection and identification uses a set of changes in the values of some parameters, to detect and identify a component malfunction. The task of assigning such sets of changes to machine status is one very much suited to ANN. Neural networks, with their remarkable ability to derive meaning from complicated or imprecise data can be used to extract patterns and detect trends that are too complex to be noticed by either humans or other computer techniques.

There are various neural network models. Among all different neural networks, the back-propagation and the probabilistic neural nets are the architectures, which have mostly been investigated for gas turbine diagnostics. The majority of the researchers refer to performance diagnostics [28, 29], while fewer refer to sensor fault detection and isolation. Kanelopoulos et al. [30] studied the performance of back-propagation (BP) neural nets for both sensor and actual engine component faults for a single shaft industrial gas turbine. The BP neural networks, however, have two main limitations: (1) difficulty of determining the network structure and the number of nodes; (2) slow convergence of the training process.

Probabilistic Neural Networks (PNN), exhibit certain advantages that make them attractive, a significant one being that their particular structure does not require a training procedure, needed for other types of neural networks. The training information is produced during the network set-up and is then embedded in its structure. PNN `training' can thus be considered to be much faster than for other types of network, such as back-propagation. Additionally, PNNs perform a probabilistic rather than a deterministic diagnosis, something closer to physical reality.

Probabilistic Neural Networks (PNN)

The Probabilistic Neural Network (PNN) is a multi-layer feed forward network. The learning procedure of this network is a supervised learning procedure. During the learning procedure the PNN classifies the training patterns to classes (represented by the output nodes). When an unknown pattern is presented to the PNN, the network estimates the probability that this pattern belongs to each class. The procedure followed and the network itself is briefly described in the following:

Let us suppose that, for training the PNN, we use the group of the m, n-dimensional, training patterns:

$$\mathbf{x}_j = \{ a_{1j}, \ a_{2j}, \ \ldots \ a_{nj} \}, \ j = 1, \ldots m$$

(17)

The graph of the resulting network is shown in figure 4. The PPN consists of three layers. The n nodes of the first layer represent the n-dimensional input. The m nodes of the second layer (hidden layer) represent the training patterns, while each one of the k nodes of the third (output) layer represents a class to which a pattern can be classified into.

Every node of the input layer of the PPN is linked to every node of the hidden layer. Each node of the hidden layer (representing a training pattern) is linked only to the node of the output layer that represents the class where the training pattern 'belongs'.

When a pattern $\mathbf{x} \in^m$ is given as an input to the network, the output is the probability density functions: $P(Si \mid \mathbf{x})$, i=1,...,k.

If we assume that the probability density functions, $P(\mathbf{x}|S_i)$, are Gaussian, we have:

$$P(S_i|\mathbf{x}) = \frac{P(S_i)}{P(\mathbf{x}) \cdot (2\pi)^{m/2} \cdot \sigma_i^m \cdot |S_i|} \cdot \sum_{j=1}^{n_i} \exp\left[\frac{-(\mathbf{x} - \mathbf{x}_j^{(i)})^T (\mathbf{x} - \mathbf{x}_j^{(i)})}{2\sigma_i^2}\right]$$

$$(18)$$

where, $x_j^{(i)}$ is the j-th pattern of the training set of patterns that 'belong' to class i, $|Si|=ni$ is the number of the training patterns that 'belong' to class i, σi is a smoothing parameter, $P(Si)$ is the 'a priori' probability of class S_i, and $P(x)$ a normalization factor representing the 'a priori' probability of patternx, which is constant assuming mutually exclusive classes, covering all possible situations.

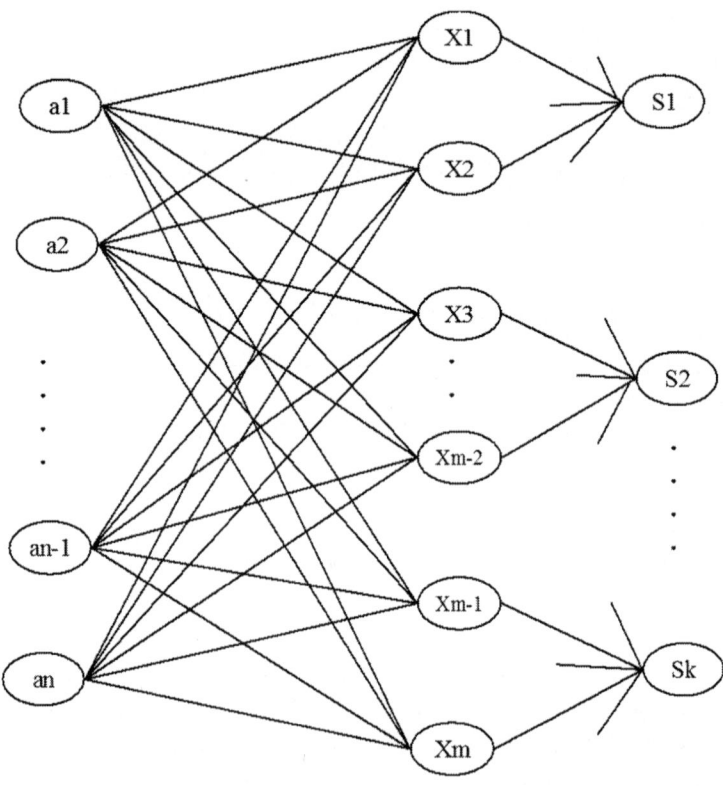

Figure 4: The general structure of the Probabilistic Neural Network.

For example if it is considered that the 'a priori' probability is equal for all classes,

$$P(S_i) = \frac{1}{k}, \quad i = 1, \ldots, k$$

(19)

During the training of the PNN, we provide the training patterns and the classes they belong to. From this information the number of nodes of each layer, as well as the links of the network with the related weights, are specified.

The weight of the link from node j1 of the input layer to node j2 of the hidden layer is:

$$w^{(1)}_{j1,j2} = a_{j1j2}$$

(20)

while, the weight of the link from node X_i of the hidden layer to node S_j of the hidden layer, is:

$$w^{(2)}_{X_i,S_j} = \frac{1}{2 \cdot \sigma_j^2}$$

(21)

where, σ_j is the smoothing parameter of class j, represented by node S_j of the output layer of the network. During the testing of the network, the probability density functions for each class are calculated, using equation (18).

Comparative and parametric investigations of the diagnostic ability of PNN on turbofan engines have been carried out in [31]. The work has also provided some general information about PNN diagnostic ability. The use of probabilistic neural networks for sensor fault detection and estimation of the sensor bias has been demonstrated in [32]. The technique proposed was shown to provide a powerful sensor validation tool, for cases where a rather limited number of measuring sensors is available, such as when data from an engine on-board an aircraft are available.

Expert Systems

In contrast to neural networks, which learn knowledge by training on observed data with known inputs and outputs, Expert systems(ES) utilize domain expert knowledge in a computer program with an automated inference engine to perform reasoning for problem solving. Three main reasoning methods for ES used in the area of engine diagnostics are rule-based reasoning, case-based reasoning and model-based reasoning. In condition monitoring practice, knowledge from domain specific experts is usually inexact and reasoning on knowledge is often imprecise. Therefore, measures of the uncertainties in knowledge and reasoning are required for ES to provide more robust problem solving. Commonly used uncertainty measures are probability, fuzzy member functions in fuzzy logic theory and belief functions in belief networks theory. An expert system dealing with uncertainty and proved to be very efficient in fault diagnosis is described below.

Bayesian Belief Network (BBN)

BBN is a probabilistic expert system, graphically represented by a set of 'nodes' and a set of 'links' connecting them. The topological features of a BBN that must be fully specified in order the network to be complete are the following: Nodes express the parameters of the represented domain. In figure 5 an example of a belief network referred to a gas turbine is presented. This network has four nodes expressing the parameters of the engine taken into account. These are: the 'efficiency factor of the high pressure compressor' (n(HPC)), the 'efficiency factor of the high pressure turbine' (n(HPT)), the 'pressure ratio' (π_c) and the 'turbine inlet temperature' (TIT).

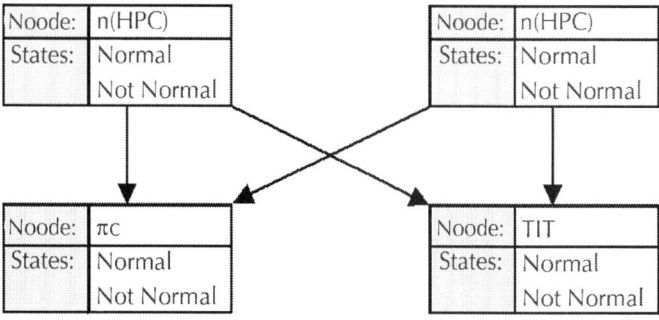

Figure 5: An example of a belief network of a gas turbine.

Each node has two or more discrete states, expressing all the different states of the parameter they refer to. For instance, in the network of figure 5, node TIT has two states: 'normal temperature' and 'not normal temperature'. In each case, the set of states of a node must be exhaustive and mutually exclusive. In other words, any possible condition of a parameter expressed by a node in a BBN is represented by one and only one state of this node. Links among the nodes express the 'rules' of interdependence that hold among them. For example, the link from node n(HPC), on the network of figure 5, to node πc expresses the fact that the condition (state) of node n(HPC) affects directly the condition of node πc. The absence of a link between two nodes doesn't mean that these two nodes are independent, but expresses the fact that the condition (state) of the one doesn't directly affect the condition of the other.

Each node has a Conditional Probability Table (CPT), expressing the probability each state of the node to occur, when the state of each other node, ending up directly to it (called 'parent node'), is known. In case that, a node has no other nodes ending up directly to it (called a 'root node'), the CPT of this node express the 'a priori' probability each state of this node to occur. In Table 1 an example of how the CPTs, of the nodes of the network of figure 5, could be, is shown.

Table 1: An example of the CPTs of the nodes of the network of figure 5

Node:	n(HPC)	
States:	Normal	Not Normal
a priori probabilities	0.90	0.10

Node:	n(HPT)	
States:	Normal	Not Normal
a priori probabilities	0.87	0.13

Node	"It		n(HPC)	n(HPT)	Parent Nodes
States	Normal	Not Normal			
Conditional Probabilities	0.97	0.03	Normal	Normal	States of parents
	0.68	0.32	Normal	Not Normal	
	0.26	0.74	Not Normal	Normal	
	0.02	0.98	Not Normal	Not Normal	

Node	TIT		n(HPC)	n(HPT)	Parent Nodes
States	Normal	Not Normal			
Conditional Probabilities	0.97	0.03	Normal	Normal	States of parents
	0.16	0.84	Normal	Not Normal	
	0.38	0.62	Not Normal	Normal	
	0.08	0.92	Not Normal	Not Normal	

Once a BBN is constructed, inference can be realized any time evidence is available. Inference is the procedure where the probabilities of each state of each node of the network are updated each time that evidence is available. 'Evidence' is the knowledge of the state of one or more nodes of the network.

Bayesian Belief Networks have some features that make them very attractive in the field of diagnosis of faults in gas turbines. The most important of these features are: BBN allow probabilistic diagnosis; it is more realistic to make diagnosis expressing the belief (probability) of whether an event occurred or not, than expressing a deterministic answer. Mathematical relationships among the variables of a network are not required in order to form a BBN. Only the way that these variables affect each other is required. This is very helpful since such mathematical relationships may be unknown. Modern approximate algorithms for inference with BBN are able, nowadays, to answer queries, once 'evidence' is provided, within few seconds, even for complicated networks, performing with adequate accuracy. Each node of a BBN can be an 'evidence' as well as a 'query'. There is no restriction to the number of 'query' or 'evidence' nodes. Therefore, there is no limitation on how many or which are the 'evidence' nodes in order to estimate the probabilities of all the other nodes of a network. It allows also the inclusion of information of different nature and from different sources for diagnostics.

Such networks have been employed in the field of gas turbine diagnostics by few researchers. Breese et al. [33], presented a method for detecting specific faults on large gas turbines that combines a thermodynamic model of the engine under examination and a BBN, constructed by use of statistical data of the engine. Palmer [34], presented a statistically also constructed BBN for fault detection of the CF6 family of engines.

The first attempt to propose a general procedure of building a BBN for diagnostic purposes, has been presented by Romessis et al. [35]. The objective of the investigation was to reveal a possible way of setting up such a network with aid of an engine performance model. The way of building diagnostic BBNs, allowing implementation into any type of engine, and the disengagement of the BBN from hard to find statistical data, were two elements that made the work interesting and promising. The effectiveness of the proposed diagnostic method was examined on benchmark fault case scenarios, in a typical modern turbofan engine of civil aviation. The diagnosis was based on the observation of fewer measurements (7) than the considered fault parameters (11). Inference with BBN showed that such a network is very reliable, since in the 96% of the cases where a fault was detected, it was detected correctly. Only a 4% of the cases were attributed to a wrong fault.

A more efficient method even in fault cases with smaller health parameters' deviations was proposed in [36]. The improvement was due to the way the BBN is constructed: probabilistic relationships among variables are more accurately represented. The effectiveness of the proposed method has been demonstrated by its strong diagnostic ability with various fault scenarios and cases at several operating conditions, including coverage of an operational envelope of a typical flight.

HYBRID AND FUSION INFORMATION TECHNIQUES

Despite research in various methods for engine fault diagnostics, there is still no method which can effectively address all issues. One way to approach the problem is to try and offset the limitations of one technique with the strength of the other. Hybrid models have attempted to bridge this gap.

An integrated fault diagnostics model for identifying shifts in component performance and sensor faults using Genetic Algorithm and Artificial Neural Network was presented in [37]. The diagnostics model operates in two distinct stages. The first stage uses response surfaces for computing objective functions to increase the exploration potential of the search space while easing the computational burden. The second stage uses concept of a hybrid diagnostics model in which a nested neural network is used with genetic algorithm to form a hybrid diagnostics model. The nested neural network functions as a pre-processor or filter to reduce the number of fault classes to be explored by the genetic algorithm based diagnostics model. The hybrid model improves the accuracy, reliability and consistency of the results obtained. In addition significant improvements in the total run time have also been observed. Ecstase [38], presents an example of the use of fuzzy logic combined with influence coefficients applied to engine test-cell data to diagnose gas-path related performance faults. The diagnostic process to identify module level engine performance faults has been validated using eight examples from real-world test-cell data. Many combinations of faults were examined in an attempt to explain the performance degradation observed in the engine under-going repair. This aspect of the process enabled the status of 17 faults to be determined, despite only five engine parameters being used. The method correctly identified the faults for all except for one fault which had a very small degradation effect on the engine performance.

A diagnostic method consisting of a combination of Kalman filters and Bayesian Belief Network (BBN) is presented in [39]. A soft-constrained Kalman filter uses a priori information derived by a BBN at each time step, to derive estimations of the unknown health parameters. The resulting algorithm has improved identification capability in comparison to the stand-alone Kalman filter. Besides the improvements in accuracy and stability, this kind of method allows information or sensor fusion, which is a very important field of research for future works. The key advantage of combining methods is that it replaces the problem of comparing classification techniques to regression techniques by the problem of choosing which information they can share. Romessis et al. [40] proposed a statistical processing of the diagnostic conclusions provided by a least-square based gas path diagnostic method, in order to improve diagnosis. In a similar attempt (see [41]) a combinatorial approach (statistical evaluation of least squares estimations) combined with fuzzy logic rules to calculate fault probabilities. The possibility of creating a mixed fault classification that incorporates both model-based and data driven fault classes was investigated in [42]. Such a classification combines a common diagnosis with a higher diagnostic accuracy for the data-driven classes. The performed analysis has revealed no limitations for realizing a principle of the mixed classification in real monitoring systems.

Information Fusion is the integration of data or information from multiple sources, to achieve improved accuracy and more specific inferences than can be obtained from the use of a single sensor alone. It is generally believed that an ensemble of methods improves diagnostic accuracy when compared to individual methods. In [43] several fusion architectures and classifiers were evaluated. Fusing classifiers that are performing very well had little positive effect. However, it was shown that fusing marginal classifiers can increase the diagnostic performance substantially, while reducing their variability. Enhanced fault localization using probabilistic fusion with gas path analysis algorithms is referred in [44],while a fusion technique allowing the merge of conclusions provided by diagnostic methods that act independently for the detection of gas turbine faults is described in [45]. The proposed technique adopts the principles of Dempster-Schafer theory for the fusion of two diagnostic methods namely a Bayesian Belief Networks (BBN) and a Probabilistic Neural Networks (PNN). The technique has been applied for the detection of thermodynamic as well as mechanical faults on gas turbines. In all cases, the effectiveness of the proposed fusion technique demonstrated that the merge of diagnostic information from different sources leads to better and safer diagnosis.

A fusion method that utilizes performance data and vibration measurements for gas turbine component fault identification is presented in [46]. The proposed method operates during the diagnostic processing of available data (process level) and adopts the principles of certainty factors theory. Both performance and vibration measurements are analyzed separately, in a first step, and their results are transformed into a common form of probabilities. These forms are interwoven, in order to derive a set of possible faulty components prior to deriving a final diagnostic decision. Then, in the second step, a new diagnostic problem is formulated and a final set of faulty health parameters are defined with higher confidence. In the proposed method the non-linear gas path analysis is the core diagnostic method, while information provided by vibration measurements trends is used to narrow the domain of unknown health parameters and lead to a well-defined solution. Finally a comprehensive presentation of different fusion possibilities offered is given in [47].

ECMD INTEGRATED SYSTEMS

Although many diagnostic methods have been proposed and some of them have been tested in real engines only few are known to be incorporated in ECMD integrated systems. An industrial monitoring and diagnostic system must comply with several requirements. For such a system to be effective it should:

- Be as automated as possible and integrated namely performing all actions from data collection to derivation of diagnostic decisions.
- Be "robust", namely not very susceptible to noise or faulty input information.
- Have an as wide as possible coverage of detectable faults. Additionally, it should allow additions of other newly discovered faults, which have not been included in the initial repertory of the system.
- Have prognostic capabilities concerning future maintenance and repair actions. This helps in ensuring that long lead-time spares are available and that outages be minimized.
- Derive information with high confidence. In this respect, derivation of the same conclusion by different methods is a very useful feature.
- Employ as few instruments as possible. The instrumentation should be kept as simple as possible and include the minimum number of instruments.
- Be modular and flexible with open circuit architecture in order to be adapted to operator's needs.
- Be very user friendly, so that it can be used by non-specialized personnel, while its output is clear enough to need very little or no interpretation.

In order to materialize a monitoring system, which possesses these features, the procedures, which should be implemented, are as follows:

- Measurement data acquisition.
- Data evaluation in order to discard unreliable readings and possibly detect sensor faults.
- Data processing using appropriate techniques in order to derive diagnostic information.
- Diagnostic inference in order to decide what is the nature, the location and the severity of a malfunction present, if any.
- Data management in order to keep historical data records for long term monitoring, without storing too much unnecessary information.

Such an integrated system and experience gained from its implementation on an operating industrial gas turbine has been presented in [48]. The main functions of this system materializing the procedures mentioned above are as follows:

Data Acquisition and Management: Data are acquired from a number of different measuring instruments, for slowly or fast varying quantities. The obtained measurements are being on-line validated and then organized in a database. The system also gives the possibility to play back measurements database in order to recreate real time operation. Additional features of the developed data acquisition feature are its flexibility and its capability to easily meet the requirements of any particular implementation.

Performance Analysis: The acquired thermodynamic measurements are being on-line processed using the adaptive modeling method [19]). Thus, at any given operating conditions, the overall engine performances and individual components health indices are being evaluated. The method can also be used off-line for the analysis of previously recorded data.

EGT Monitoring: The hot section, being the most critical area of the engine, is receiving particular attention, through exhaust gas temperature profile monitoring ([49]. This monitoring provides indication of possible burner malfunctions or thermocouple faults. Off- line analysis of historic data stored in measurements database can also be performed.

Vibration Monitoring: The means of identifying mechanical faults are provided by this function of the system. For data from vibration sensors, the following diagnostic features are extracted and assessed: a) overall vibration level, b) power spectra (on-line frequency analysis), and c) spectral signatures [50]). Finally, as the other monitoring modules, it offers the possibility of off-line line analysis of historic data stored in measurements database.

These functions are performed by the system continuously, while the engine is in operation. Their implementation provides adequate diagnostic information about engine condition. This information is being further assessed using a rule based inference engine that provides an engine condition assessment. Thus, the user is being informed in real time about the engine's condition and performance. The main interface of the system implemented on a PC is shown in figure 6a. It comprises an axial cut-out of the monitored engine and gives the most critical information about the engine condition. The system offers the possibility to perform a more detailed analysis by activating the previously described functions through the buttons on the menu at the upper right hand corner.

An example of system effectiveness in diagnosing is the following. A twin-shaft industrial gas turbine with 21 MW nominal output, used for electricity production in a power station, is considered. The turbine suffered from the formation of deposits on gas generator and power turbine blades, very soon after it was put on operation (see figure 1a). A remedy action taken by the manufacturer was a small re-staggering (opening) of power turbine stationary blades. An easy and reliable way of identification of the malfunction of the turbine is provided by the method of adaptive modeling. The technique has been applied to test data from this turbine and it gave a clear picture of the problem. Comparison of health parameters deviation obtained from data from the initial condition of the engine and after the presence of the problem was detected is shown in figure 6b. It is clearly shown that the swallowing capacity of both turbines has been significantly reduced, as factor f3 shows a reduction of more than 1.5% and f5 more than 3%. The reduction in f1 (of ~ 0.8%) indicates that the compressor has also suffered some deterioration.

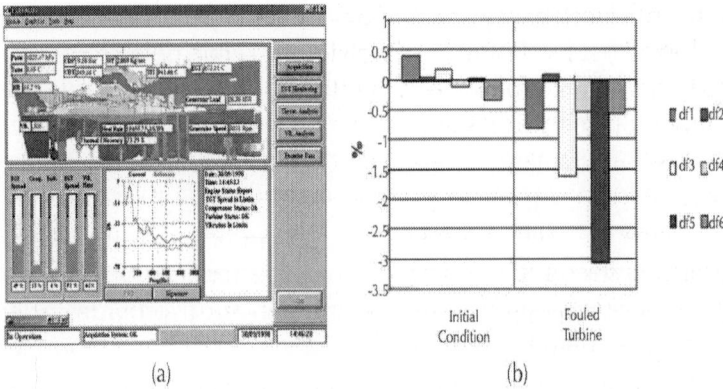

(a)   (b)

Figure 6: (a) Display of a user friendly monitoring software for an industrial gas turbine. (b) Health Indices Percentage deviation, for a gas turbine, which has suffered severe turbine fouling, caused by fuel additives.

CONCLUSIONS

In this chapter, we have attempted to present basic principles of the engine condition monitoring and diagnostics (ECMD) subject. It would be impossible to cover in few pages all the aspects of ECMD. Thousands of papers have been published and a vast amount of knowledge has been accumulated. Even extensive reviews cannot mention all the proposed methods. In this respect we presented selective methods representative of three main steps of an ECMD approach, namely data acquisition, data processing and diagnostic decision-making, with emphasis on the last two steps. Few recently developed hybrid, data and method fusion techniques have also been briefly discussed. The structure of an integrated ECMD system incorporating different diagnostic technics and already in operation is also presented.

The following conclusions are the outcome of over twenty five years of experience in the area of ECMD.

- The main problems with respect to the industry adoption of advanced technics are the following: a) lack of data due to no data collection and/or data storage at all; b) lack of efficient communication between method developers and maintenance practitioners; c) lack of efficient validation approaches.

- Both physics based and data-driven models show benefits and drawbacks. From the decision making point of view both traditional and artificial intelligence techniques are used, although it seems that hybrid approaches are more promising.
- The value of vibration monitoring and other sources data in refining gas-path monitoring results has been recognized. The approach of combining different monitoring results, i.e., data fusion, is becoming an active area of research.
- Usage and life monitoring for fatigue critical or life-limited parts are increasingly important
- Collaboration of ECMD research groups is necessary in order to produce integrated platforms for enhancing an ECMD system since each research group has its own specialty and focus in the area.

The following research directions are required for the next generation of ECMD systems: Enhancement of ECMD systems to collect accurate information, especially fault event information. This information would be very useful for model building and model validation as well. Advanced models and methods for utilization of the transient data diagnostic information as well as detailed higher order models for deterioration mechanisms and faults reliable simulation should be developed. Accurate prognostic models development is also necessary. Finally, there is a need for establishment of efficient validation approaches through benchmark test cases to compare the merits and the drawbacks of different modeling and algorithmic approaches.

REFERENCES

1. Zaita AV, Buley G, Karlsons G. Performance deterioration modeling in aircraft gas turbine engines. Journal of Engineering for Gas Turbines and Power. 1998;120(2):344-9.

2. Kurz R, Brun K. Degradation in gas turbine systems. Journal of Engineering for Gas Turbines and Power. 2001;123(1):70-7.

3. Diakunchak IS. Performance Deterioration in Industrial Gas-Turbines. Journal of Engineering for Gas Turbines and Power. 1992;114(2):161-8.

4.   Li YG. Performance-analysis-based gas turbine diagnostics: A review. Proceedings of the Institution of Mechanical Engineers, Part A: Journal of Power and Energy. 2002;216(5):363-77.

5.   Marinai L, Probert D, Singh R. Prospects for aero gas-turbine diagnostics: A review. Applied Energy. 2004;79(1):109-26.

6.   Urban LA, Volponi AJ. Mathematical methods of relative engine performance diagnostics. SAE 1992 transactions, vol. 101, journal of aerospace, technical paper 922048.

7.   Barwell MJ. Compass - Ground Based Engine Monitoring Program for General Application. 1987. SAE technical paper 871734.

8.   Doel DL. TEMPER - a gas-path analysis tool for commercial jet engines. Journal of Engineering for Gas Turbines and Power. 1994;116(1):82-9.

9.   Doel DL. An assessment of weighted-least-squares-based gas path analysis. J Eng Gas Turbines Power, Trans ASME. 1994;116:365-73.

10. Stamatis A, Papailiou KD. Discrete operating condition gas path analysis. AGARD-CP-448, Engine Condition Monitoring - Technology and Experience. 1988.

11. Gronstedt TV. A multi point gas path analysis tool for gas turbine engines with a moderate level of instrumentation. 2001.XV ISABE, Bangalore, India, Sept. 3-7.

12. Gulati A, Taylor D, Singh R. Multiple operating point analysis using genetic algorithm optimization for gas turbine diagnostics. 2001. XV ISABE, Bangalore, India, Sept. 3-7

13. Pinelli M, Spina PR, Venturini M. Optimized Operating Point Selection for Gas Turbine Health State Analysis by Using a Multi-Point Technique. ASME Turbo Expo 2003, collocated with the 2003 International Joint Power Generation Conference (GT2003) June 16–19, 2003, Atlanta, Georgia, USA, Paper no. GT2003-38191.

14. Mathioudakis K, Kamboukos P. Assessment of the effectiveness of gas path diagnostic schemes. Journal of Engineering for Gas Turbines and Power. 2006;128(1):57-63.

15. Gronstedt T. Identifiability in multi-point gas turbine parameter estimation problems. ASME Turbo Expo 2002: Power for Land, Sea, and Air (GT2002) June 3–6, 2002, Amsterdam, The Netherlands, Paper no. GT2002-30020.

16. Henriksson M, Borguet S, Leonard O, Gronstedt T. On inverse problems in turbine engine parameter estimation.. ASME Paper no. GT2007-27756

17. Aker GF, Saravanamuttoo HIH. Predicting gas turbine degradation due to compressor fouling using computer simulation techniques. ASME paper 88-GT-206.

18. Stamatis AG. Evaluation of gas path analysis methods for gas turbine diagnosis. Journal of Mechanical Science and Technology. 2011;25(2):469-77.

19. Stamatis A, Mathioudakis K, Papailiou KD. Adaptive simulation of gas turbine performance. Journal of Engineering for Gas Turbines and Power. 1990;112(2):168-75.

20. Li YG, Korakianitis T. Nonlinear weighted-least-squares estimation approach for gas-turbine diagnostic applications. Journal of Propulsion and Power. 2011;27(2):337-45.

21. Ogaji S, Sampath S, Singh R, Probert D. Novel approach for improving power-plant availability using advanced engine diagnostics. Applied Energy. 2002;72(1):389-407.

22. Zedda M, Singh R. Gas turbine engine and sensor fault diagnosis using optimization techniques. Journal of Propulsion and Power. 2002;18(5):1019-25.

23. Mathioudakis K, Kamboukos P, Stamatis A. Gas turbine component fault detection from a limited number of measurements. Proceedings of the Institution of Mechanical Engineers, Part A: Journal of Power and Energy. 2004;218(8):609-18.

24. Grodent M, Navez A. Engine Physical Diagnosis Using a Robust Parameter Estimation Method., 37th AIAA/ASME /SAE/ASEE Joint Propulsion Conference and Exhibit, 8-11 July 2001, Salt Lake City, Utah, paper AIAA-2001-3768.

25. Mathioudakis K, Kamboukos P, Stamatis A. Turbofan performance deterioration tracking using nonlinear models and optimization techniques. Journal of Turbomachinery. 2002;124(4):580-7.

26. Stamatis A, Mathioudakis K, Papailiou K. Optimal measurement and health index selection for gas turbine performance status and fault diagnosis. Journal of Engineering for Gas Turbines and Power. 1992;114(2):209-16.

27. Ogaji SOT, Sampath S, Singh R, Probert SD. Parameter selection for diagnosing a gas-turbine's performance-deterioration. Applied Energy. 2002;73(1):25-46.

28. Eustace R, Merrington G. Fault diagnosis of fleet engines using neural networks. Fault Diagnosis of Fleet Engines Using Neural Networks. ISABE paper, ISABE 95-7085.

29. Volponi AJ, DePold H, Ganguli R, Daguang C. The use of kalman filter and neural network methodologies in gas turbine performance diagnostics: A comparative study. Journal of Engineering for Gas Turbines and Power. 2003;125(4):917-24.

30. Kanelopoulos K, Stamatis A, Mathioudakis K. Incorporating neural networks into gas turbine performance diagnostics. ASME paper 97-GT-35.

31. Romessis C, Stamatis A, Mathioudakis K. A parametric investigation of the diagnostic ability of probabilistic neural networks on turbofan engines. ASME paper 2001-GT-0011.

32. Romesis C, Mathioudakis K. Setting up of a probabilistic neural network for sensor fault detection including operation with component faults. Journal of Engineering for Gas Turbines and Power. 2003;125(3):634-41.

33. Breese JS, Gay R, Quentin GH. Automated decision-analytic diagnosis of thermal performance in gas turbines. ASME paper 92-GT-399.

34. Palmer CA. Combining Bayesian belief networks with gas path analysis for test cell diagnostics and overhaul. ASME paper 98-GT-168.

35. Romessis A, Stamatis A, Mathioudakis K. Setting up a belief network for turbofan diagnosis with the aid of an engine performance model. ISABE-2001-1032.

36. Romessis C, Mathioudakis K. Bayesian network approach for gas path fault diagnosis. Journal of Engineering for Gas Turbines and Power. 2006;128(1):64-72.

37. Sampath S, Singh R. An integrated fault diagnostics model using genetic algorithm and neural networks. Journal of Engineering for Gas Turbines and Power,2006;128, (1):49-56.

38. Eustace RW. A real-world application of fuzzy logic and influence coefficients for gas turbine performance diagnostics. Journal of Engineering for Gas Turbines and Power. 2008;130(6).

39. Dewallef P, Romessis C, Leonard O, Mathioudakis K. Combining classification techniques with Kalman filters for aircraft engine diagnostics. Journal of Engineering for Gas Turbines and Power. 2006;128(2):281-7.

40. Romessis C, Kamboukos P, Mathioudakis K. The use of probabilistic reasoning to improve least squares based gas path diagnostics. Journal of Engineering for Gas Turbines and Power. 2007;129(4):970-6.

41. Lipowsky H, Staudacher S, Nagy D, Bauer M. Gas turbine fault diagnostics using a fusion of least squares estimations and fuzzy logic rules. ASME Turbo Expo 2008: Power for Land, Sea, and Air (GT2008)June 9–13, 2008, Berlin, Germany. ASME Paper no. GT2008-50190

42. Loboda I, Yepifanov S. A mixed data-driven and model based fault classification for gas turbine diagnosis. Proceedings of ASME Turbo Expo 2010: International Technical Congress, 8p., Scotland, UK, June 14-18, Glasgow, ASME Paper No. GT2010-23075.

43. Donat W, Choi K, An W, Singh S, Pattipati K. Data visualization, data reduction and classifier fusion for intelligent fault diagnosis in gas turbine engines. Journal of Engineering for Gas Turbines and Power. 2008;130(4).

44. Kyriazis A, Mathioudakis K. Enhanced fault localization using probabilistic fusion with gas path analysis algorithms. Journal of Engineering for Gas Turbines and Power. 2009;131(5).

45. Romessis C, Kyriazis A, Mathioudakis K. Fusion of gas turbines diagnostic inference - The dempster-schafer approach. Proceedings of IGTI/ASME Turbo Expo 2007, 9p., Canada, May 14-17, 2007, Montreal, ASME Paper GT2007-27043.

46. Kyriazis A, Tsalavoutas A, Mathioudakis K, Bauer M, Johanssen O. Gas turbine fault identification by fusing vibration trending and gas path analysis. ASME Turbo Expo 2009: Power for Land, Sea, and Air (GT2009),June 8–12, 2009, Orlando, Florida, USA, ASME Paper no. GT2009-59942

47. Volponi AJ, Brotherton T, Luppold R. Development of an information fusion system for engine diagnostics and health management. AIAA 1st Intelligent Systems Technical

Conference20 - 22 September 2004, Chicago, Illinois, AIAA Paper 2004-6461.

48.   Mathioudakis K, Stamatis A, Tsalavoutas A, Aretakis N. Performance analysis of industrial gas turbines for engine condition monitoring. Proceedings of the Institution of Mechanical Engineers, Part A: Journal of Power and Energy. 2001;215(2):173-84.

49.   Tsalavoutas A, Mathioudakis K, Smith MK. Processing of circumferential temperature distributions for the detection of gas turbine burner malfunctions. ASME Paper 96-GT-103.

50.   Loukis E, Wetta P, Mathioudakis K, Papathanasiou A, Papailiou A. Combination of different unsteady quantity measurements for gas turbine blade fault diagnosis. ASME paper, 91-GT-201, International Gas Turbine and Aeroengine Congress and Exposition, June 3-6 1991, Orlando.

Citations

CHAPTER 1

Rainer Kurz, Matt Lubomirsky, and Klaus Brun, "Gas Compressor Station Economic Optimization," International Journal of Rotating Machinery, vol. 2012, Article ID 715017, 9 pages, 2012. doi:10.1155/2012/715017.

CHAPTER 2

Anand Kumar, Vladimir Grupcev, Meryem Berrada, Joseph C Fogarty, Yi-Cheng Tu,Xingquan Zhu, Sagar A Pandit, and Yuni Xia, DCMS: A data analytics and management system for molecular simulation, doi:10.1186/s40537-014-0009-5.

CHAPTER 3

Stefano Bianchi, Alessandro Corsini, Anthony G. Sheard, and Cecilia Tortora, "A Critical Review of Stall Control Techniques in Industrial Fans," ISRN Mechanical Engineering, vol. 2013, Article ID 526192, 18 pages, 2013. doi:10.1155/2013/526192.

CHAPTER 4

Han Li and De-yun Xiao, Fault Diagnosis of Tennessee Eastman Process Using Signal Geometry Matching Technique, doi: 10.1186/1687-6180-2011-83.

CHAPTER 5

Krzysztof Biernat (2015). The Influence of Engine Fuel Manufacturing Processes on Their Performance Properties in Operating Conditions, Storage Stability of Fuels, Prof. Krzysztof Biernat (Ed.), ISBN: 978-953-51-1734-6, InTech, DOI: 10.5772/59800.

CHAPTER 6

Krzysztof Biernat (2015). Criteria for the Quality Assessment of Engine Fuels in Storage and Operating Conditions, Storage Stability of Fuels, Prof. Krzysztof Biernat (Ed.), ISBN: 978-953-51-1734-6, InTech, DOI: 10.5772/59801.

CHAPTER 7

O. Velde, G. Kreuzfeld and I. Lehmann, "Reverse Engineering of Turbocharger Compressor Designs Based on Non-Parametric CAD Data," Engineering, Vol. 4 No. 7, 2012, pp. 353-358. doi: 10.4236/eng.2012.47046.

CHAPTER 8

Anastassios G. Stamatis (2013). Engine Condition Monitoring and Diagnostics, Progress in Gas Turbine Performance, Dr. Ernesto Benini (Ed.), ISBN: 978-953-51-1166-5, InTech, DOI: 10.5772/54409.

Index